EXPLORING OCEAN FRONTIERS

A Background Book on Who Owns the Seas

EXPLORING OCEAN FRONTIERS

A Background Book on Who Owns the Seas

by *Frances and Walter Scott*

With a Preface by Jacques Piccard

PARENTS' MAGAZINE PRESS • NEW YORK

Each Background Book is concerned with the broad spectrum of people, places and events affecting the national and international scene. Written simply and clearly, the books in this series will engage the minds and interests of people living in a world of great change.

To Norman Levy and Joseph Vollmerhausen

Copyright © 1970 by Frances and Walter Scott
All rights reserved
Printed in the United States of America
International Standard Book Numbers: Trade 0-8193-0321-6, Library 0-8193-0322-4
Library of Congress Catalog Card Number: 71-107231
Design by Patricia Ayearst

Acknowledgments

We have used a number of books in our research that seem too technical or too specialized to be included in "Suggested for Further Reading" (pages 205 to 208), as follows: The standard work on the Law of the Sea is *Public Order of the Oceans* by Myres McDougal and William Burke (New Haven: Yale, 1962). The complete text of the 1958 Geneva Conventions appears in that book. *International Law of the Sea* by C. J. Colombos (London: Longmans, Green, 1962) traces past history of the law from a British viewpoint. Two books on fisheries are: *International Law of Fisheries* by Douglas M. Johnson (New Haven: Yale, 1965) and *Common Wealth in Ocean Fisheries* by Francis Christy and Anthony Scott (Baltimore: Johns Hopkins, 1965). Three newsletters were very useful: *Ocean Science News, Oceanology,* and *Underwater Letter.*

We would like to thank the following persons who provided technical advice and read the book in manuscript: Leo Geyer and James Wells of Grumman Aerospace; William Fuller, a maritime lawyer; and Edward Schaeffers of the Bureau of Commercial Fisheries. Beatrice Solensky Burke was most helpful with the typing. The drawings in the text were done by Bernard Parke.

FRANCES AND WALTER SCOTT

Centerport, N. Y.
1970

Contents

Acknowledgments	v
Introduction: To Whom Does the Sea Belong?	ix
1 The Changing Sea	1
2 The Historic Sea	19
3 The Three-Dimensional Sea	40
4 Oil and Other Minerals	65
5 Ocean Fisheries	88
6 The Technological Revolution: Submersibles	111
7 The Technological Revolution: Habitats	133
8 The Uses of the High Seas	154
9 The Future of the Sea	173
Documents:	
I Excerpts from the 1958 Geneva Conventions	191
II The Truman Proclamation	198
III The Bartlett Bill	200
Notes	203
Suggested for Further Reading	205
Index	209
Picture Credits	220

To Whom Does the Sea Belong?

One day, in the spring of the year 200,000 B.C., a group of children were playing on a beach. The sea was noticeably calm that day. The children belonged to a tribe that lived close to an estuary. Amidst an accumulation of floating debris which drifted down the river, seagulls, emitting their raucous cries, flew, fished repeatedly, landed on the water and took off again. From time to time one or another of them perched on a tree trunk floating near the river bank. Suddenly one of the children noticed one of these seagulls and pointed it out to his companions. His primitive mind marveled that the seagull could float on a log. He approached the seagull, watched it fly away, studied the tree trunk, and in a flash of inspiration, despite the cries of his terror-stricken playmates, this son of *Homo sapiens* jumped on the tree trunk and discovered that he, too, could be carried afloat. He had invented the boat. That day, without a doubt, the sea, all of the sea, belonged entirely to him alone.

Two hundred thousand and 1,962 years later, a Russian ship was sailing to Cuba, loaded with missiles which could menace the independence and integrity of the United States. A group of American warships, backed by intense diplomatic activity, put to sea and traveled toward the intruder. The latter made an about-face. At that instant, no one on earth doubted that the sea belonged to the United States.

The nation that is backed by resourcefulness or by the force of numbers and superior strength can maintain possession of the sea by the simple right of occupancy and by that of might. This harsh law, humiliating for those who believe in justice and censure the theory of force, is sometimes glossed over, embellished by international treaties that are more or less respected, but which only work out when acting in the interest of the nations that sign them, and indeed when supported by the power of those nations.

This power is not always measured directly by canons or atomic bombs. It can be measured by the number of voices in an international assembly, but has not the number of voices itself been established beforehand as a function of the power of the respective countries, or simply, by that of oil deposits, or geographic situations, or of all the other economic interests, as strong, after all, as an army?

Whenever the more numerous countries of the world are seated around the conference table of the United Nations, a certain tendency is evident; that is, that this international assembly should have the ability to manage "peacefully" all the common treasures of our planet, particularly those hidden in the sea. Since laws are meaningless without provision for execution, it would be necessary to establish a strong executive system to enforce these laws. Who will have this responsibility? If the Security Council has the responsibility, then it is in the hands of the five great powers holding the right to veto; if the General Assembly has the responsibility, then it belongs to the majority, who find themselves to be a group of nations having nothing, or almost nothing invested in the sea, and who have not yet participated in its exploration. Furthermore the great powers would doubtless ignore this majority if their own rights to cultivate the fruits of their own labor and effort were denied. It is evident that the gap between the idealists and the pacifists on one side, and the realists and the mighty on the other is still great, so great that no one can predict when it will be bridged.

But, in fact, what concerns them so? What do they wish to conquer? What do they want to share? What do they seek to divide among themselves? No one entirely knows the ocean; no one can pretend to have made a comprehensive survey of its fabulous riches. Before wanting to distribute it or to plunder it, would it not be best instead to inventory it, to analyze it, to dissect it, to study it? An immense field exists, which for the moment is reserved for a small minority of scientists: in their laboratories, on land, on the sea's surface, in the depths, collecting, harvesting, assembling, calculating, interpreting, learning little by little what the ocean is and how it can serve us. Others assist them, working often thanklessly, but with great merit, and convey through their analyses to those who follow, the results of their research. This task of gathering and classifying information will bear fruit in the future, and it is this work that perhaps one day will provide the specialists with the wisdom to judge and decide at that time to whom the sea belongs.

Walter and Frances Scott's book has been written in this light. It is a remarkably researched work; it has a wealth of information and references of great value that will serve precisely to enlighten those who will later work out the laws of the sea. It is a building block that will help to construct the dike that should one day protect humanity against the tide of arbitrariness and of force, that has assailed it unceasingly since its ancient beginnings, a tide that has presided since the division of the earth's continents, but perhaps which, thanks to efforts like this one, will wash ashore for the world the means to share equitably in the ocean's riches.

JACQUES PICCARD

*Translated from the French
by Edmond Rabut
December, 1969.*

CHAPTER 1

The Changing Sea

From his prehistoric beginning, man's personal relationship with the sea has been unique, unlike any other tie with his natural surroundings. Men in ancient times could claim the land, build fences around a farmer's plot, or raise stone walls to guard a city. Armies could seize, occupy, and defend territory. But who could own the ocean? It belonged to everyone and to no one. Men who cut down trees and burned fields could not control crushing waves, rising tides, or relentless currents. At best, they could sail across protected bays and catch fish along the shore. Governments might stifle landsmen, reducing some to slavery, but fishermen and sailors maintained a high degree of independence. The sea was their hiding place and avenue of escape. A source of food, a pathway, barrier, or battleground, the sea has fascinated men with its lure of freedom and danger.

This personal relationship to the sea persists in modern

times and influences the laws that control it, although new advances in science and technology in the twentieth century are rapidly breaking down the protective blanket the oceans once provided. The expanding world population, coupled with man's ability to exploit or destroy his environment, is disproving the age-old belief that the seas are inexhaustible.

New technology has encroached upon human feelings about the sea—an emotional involvement inherited from man's prehistoric ancestors, as old as—if not older than—the farmer's attachment to the soil. When a sailor today crosses the Atlantic alone in a small vessel, or the Norwegian scientist Thor Heyerdahl sails in a papyrus boat, his motives are deeper than a desire for headlines. They recall a time when men relied on courage and skill to conquer the unknown ocean.

In 1967, Francis Chichester, an Englishman, sailed his 54-foot ketch, *Gypsy Moth,* around the world alone, in 226 days, with only one stop—at Australia. In a period when commercial jets circle the globe in less than a day, Chichester was attempting to match clipper-ship records in the world's roughest and most awe-inspiring waters, the Roaring Forties of the South Pacific. He said he had abandoned airplanes for sailboats because flying had lost the attraction of a pioneer venture and had become a matter of technical training and pilot skill. Aircraft had become commonplace in the most distant parts of the world.

When Chichester faced the sea as a last frontier he found it had become quite crowded. He wrote of his rounding Cape Horn in a raging gale:

> When I stepped into the cockpit, I was astounded to see a ship nearby, about a half mile off. I had a feeling that if there

was one place in the world where I would not see a ship, it was off Cape Horn.[1]

It was a Royal Navy ice-patrol ship, H.M.S. *Protector,* with newsmen aboard to record Chichester's passage of the famous Horn.

Later, as the wind strengthened, Chichester struggled to stay alert in the dangerous seas:

I was beginning to feel seasick, and had the usual reluctance to do anything. I just wanted to be left alone, by things and especially by people. I cursed the *Protector* for hanging about, especially as she looked steady enough to play a game of billiards on her deck. Just then damned if an aircraft didn't buzz into sight. I cursed it. If there was one place in the world where I expected to be alone it was off Cape Horn . . .[2]

In spite of its dangers, men are attracted to the beauty and mystery of the sea, the challenge of its sudden, often unpredictable changes, from placid calm to bone-chilling fog or raging storm. Many feel a personal love for the sea as a source of life, rather than a cold and dangerous frontier to be conquered. Long before science developed a formal theory on the subject, men intuitively felt—perhaps from the salt water in their blood—that the oceans contain the source of life on this planet.

Rachel Carson, who devoted a lifetime to scientific study of the sea and to explaining its story to laymen, described the gray beginnings of life a billion years ago. The first living things may have been simple micro-organisms like some of the bacteria we know today; they were "not quite plants, not quite animals, barely over the intangible line that separates the non-living from the living."[3] This first life probably did not contain chlorophyl, with which plants in sunlight change chemicals into living tissue. Little sunlight could penetrate through the endless rains and the sea's few organisms lived either on organic substances in the water or directly on inorganic food.

Rachel Carson also described the epoch of mountain-building, when the earth's crust was cooling and contracting. The torrential rains were eroding the land, which still had no plant cover to protect it. Some 350 million years ago the first land life crept on shore:

> It was an arthropod, one of the great tribe that later produced crabs and lobsters and insects. It must have been something like a modern scorpion, but, unlike its descendants, it never wholly severed the ties which united it to the sea. . . . It lived a strange life, half-terrestrial, half-aquatic, something like that of the ghost crabs that speed along the beaches today, now and then dashing into the surf to moisten their gills.[4]

Millions of years later, plant life moved onto the land, and the first amphibians appeared, as fins developed into legs and gills into lungs. Some of the air-breathing mammals returned to the sea as whales, dolphins, and seals. Today human beings are completing the cycle, returning to the water—first in hard-hat diving suits, then with self-contained breathing apparatus such as the Aqua-Lung. Men are even experimenting with the idea that perhaps the lungs of mammals once again could do the work of gills. Working with mice, experimenters currently are submerging them in water for longer and longer periods to see whether normally air-breathing lungs, flooded with liquid, can be made to extract oxygen directly from the water. In the future, men may learn to breathe under the ocean just as they do on land.

While adventurers like Chichester brave the open seas, millions of ordinary people are attracted to the shore, finding both recreation and livelihood along the inshore waters, the tidal flats, estuaries, bays, and beaches that line the coast. These shallow areas, where the sun penetrates, become incubators of marine life and food sources for roaming seabirds. Here again, human beings feel a personal tie with the sea that transcends private beach-front ownership, industrial might, or government rules and regulations. The individual feels he has a right to fish, swim, dive, navigate, anchor, or cause pollution in the sea whenever he wishes. The Sunday afternoon outboarder becomes an explorer as soon as he leaves the dock. Simultaneously, however, he becomes a despoiler because his outboard exhaust leaves an oil film on the water, his indestructible plastic sandwich bags clog up other outboard intakes, and his non-corrosive aluminum beer cans leave a long-lasting record of our civilization on the ocean floor.

Yet, even beer cans have their uses. Off the coast of California, scientists have been studying the rate of sediment build-up on the ocean floor. New dams and flood-control throughout the state have prevented river sediment from reaching the sea and replenishing desirable beaches. Using a small, deep submersible, *Deepstar-4,000*, the scientists found it easy to measure sediment rates over hundreds of millions of years, but almost impossible to gauge short-term deposits without a reliable frame of reference. Beer cans were everywhere, but the scientists managed to turn this eyesore into an advantage. Knowing that flip-top beer cans had come into use only a few years earlier, they measured the sediment around these cans to determine the rate of deposit since that time. Unfortunately, not all debris can be turned to such effective scientific use.

Early man first used the sea as a source of food rather than for transportation. Primitive man fished with a spear, with traps, and with tidal fish nets. He made his fish line of vines or thongs, and his fishhooks from thorns or bone. Shellfish were abundant and required no tools other than rocks to smash them open.

Great mounds of oyster shells along the coasts of the United States provide excellent archaeological evidence about the earliest inhabitants of the land. Oyster shells have been discovered on the floor of the continental shelf off the east coast, and fishermen occasionally have brought up the bones and teeth of such prehistoric animals as the mastodon. Eleven thousand years ago the entire continental shelf was exposed for an average distance of 70 miles from land, which corresponds to a depth of approximately 200 feet below the present sea level along the United States coast. The oyster shells are believed to have come from the refuse piles of a prehistoric race that entered America from the northwest about 12,000 years ago. The outer edge of the seabed that was once above water shows shoreline characteristics similar to those of the present Atlantic beaches, marshes, and estuaries.

Before the domestication of animals, fish and shellfish were primary sources of protein for early man, more important in his diet than in ours. Nevertheless, we also depend on ocean products. Fish meal feeds the poultry that supply breakfast eggs, broilers, fryers, and Thanksgiving turkeys. Protein available from the sea is being considered as a means of feeding the world's undernourished and starving millions.

Experts, trying to gauge whether estimates of food from the sea are exaggerated, are studying both plankton supplies in the ocean and the food chain (big fish eating little fish) – the factors that eventually result in a usable food product.

Plankton – the minute plant and animal organisms floating in the water – is the basic source of food in the sea. Just as arable land on the continents is limited, ocean resources also are exhaustible. Contemporary man, however, must go a long way before his utilization of the ocean is comparable to primitive man's domestication of animals. Despite modern technology, he is still basically a hunter in the sea, and only now is he beginning to explore the idea of sea farming or animal husbandry. The English language does not even include words to describe controlled methods of increasing or improving the sea's output of food, except "aquaculture" or "mariculture," both of which are artificial and only slowly coming into use.

Early man, gathering shellfish along the beaches, found the sea at first a barrier and then a pathway open for travel. Land transportation was never easy, especially before horses came into general use. Mountains and deserts blocked the way, and hostile tribes denied access to or collected tribute from travelers. In contrast, the sea was open to all. Anyone able to build a ship could use it, provided he had the necessary skill and courage, for it was a hazardous and uncharted world. Some peoples, such as the Phoenicians, made the sea a way of life, growing wealthy on maritime trade. For the Greeks, the wine-dark sea was a pathway to commerce, conquest, and adventure.

Even the myth-memory of Western Civilization, as far back as the Greeks, is pervaded with the lure of the sea.

Jason sailed for the Golden Fleece; Theseus was taken by ship with twenty youths and maidens as tribute to the Minotaur of Crete; Agamemnon, with the combined might of the Greek states, amassed a fleet to sail to Troy. In one of the greatest hero journeys, Odysseus spent ten years of hardship and adventure on the sea before finding his way back to patient Penelope.

Travelers, confronted by the sea, felt the awe and terror reflected in legends. The unknown sea suggested death, along with the possibility of great marvels and isles of enchantment just over the horizon. The Italian poet Dante wrote that Odysseus, as an old man approaching death, sailed with a boatload of companions into the western sunset, where they were engulfed by a tempest. Similarly, King Arthur, mortally wounded, was carried by water to Avalon–probably an early name for Glastonbury, an English town surrounded by marshes, or perhaps for some Celtic Blessed Isle. From Avalon the hero was to return, the Once-and-Future King. Atlantis, the fabulous island that vanished under the water, may reflect the sea's mysterious character, or, as scientists are once again considering, may have been a mountain that sank in some prehistoric earthquake.

Behind the legends was the reality of the seas as they were used for navigation and commerce: the Mediterranean with its central position in the Ancient World; the Atlantic during the time of discovery and colonization; and the Pacific, so important during World War II. Even at the height of the Dark Ages, when land travel had almost slowed to a halt, the Norsemen used sea routes to terrorize Britain and France and to spread colonies as far as Greenland. Maritime commerce and travel reached their peak in the nineteenth century, only a few decades before their inevitable decline in the face of undersea cables, radio, aircraft, and pipelines.

If the sea carried men over great distances, it also kept them apart, dividing neighbors and protecting seagirt nations. In warfare, the difficulty of transporting heavy gear and horses by water discouraged overseas invasions, and losses from unsuccessful attempts were enormous. Invasion by water requires extensive preparation and materiel, as successful commanders throughout history—from William the Conqueror to General Eisenhower—have found out. The desire for independence in the North American colonies was fostered by the breadth of ocean separating them from the mother country, and the United States, even in the days of westward pioneering, has never been able to ignore the sea completely. It is a fact of national existence, a means of livelihood, and until World War II, a barrier against potential enemies.

In the twentieth century, uses of the sea have changed so fast that the most primitive and most sophisticated forms of sea transportation exist side by side and sometimes conflict. Jet aircraft have overcome the extended time and danger of sea travel. They cross the Atlantic in a few hours, with no need for sleeping accommodations or reserve supplies of food. Airspace above the ocean has become part of the sea, to be included in laws concerning ocean sovereignty. A man in a boat looks at the sea differently because it is no longer the single available method of travel. He can choose, by taking a jet, to avoid the endless miles of water and the empty days of sky and ocean. Although he may prefer the deep satisfaction of confronting the sea, his relationship to it has subtly changed.

As pathway or as barrier, the oceans, in the past, have been a battlefield on which men fought, either for immediate control of the water or for influence in the balance of power on land. In World War II, for instance, the whole Pacific was a battleground for the Japanese and United States fleets with their air arms, the newest and most deadly extension of sea power. At Pearl Harbor, the Japanese opened the war by trying to destroy United States naval forces in the Pacific, and one decisive factor in terminating the war was the rebuilding of the United States fleet.

But sea power has changed since World War II because of changes in submarine warfare. When submarines were introduced in World War I, they could sink or damage merchant shipping and conceal themselves fairly well from the enemy's armed vessels. By 1939 they were more effective in attack, but they continued to fight near the surface, using their ability to submerge for concealment. Now, nuclear submarines travel for long periods far beneath the surface, needing no oxygen or refueling. They carry ballistic missiles with nuclear warheads that can travel 2,500 miles.

Because of new technology, the United States, which once believed in isolation behind the protection of two oceans, has been forced to create new methods of defense. A screen of underwater listening devices guards the land against foreign submarines. Instead of concentrating on the Atlantic and Pacific only, the United States defensive weapons also face Canada to intercept missiles that might come across the Arctic.

However, only the most technically advanced nations have nuclear submarines: England, France, the United States, the Soviet Union, and possibly communist China. The complex engineering and highly developed industrial organization needed to build underwater structures or to send men to the deepest ocean are not available to countries which do not have the necessary capital to develop them.

In its familiar guise, the sea remains what it always has been, a place for small fishermen or for ore boats and tankers that transport bulk cargo unsuited to railroads or aircraft. In the exchange of goods among nations, the largest tonnage continues to move by water. Although little is reported in the press about accidents in ordinary maritime trade, hazards for ships have not disappeared. Collision, stranding, fire, and explosion take their toll every year.

Modern technology has added a new dimension to the sea, a world under the surface dimly visualized before, but never within actual reach. The conquest of ocean space may prove more vital to our survival than moon walks or Venus probes. New resources, such as oil, natural gas, minerals, and marine bottom life have become available. Small submarines from which men can observe the ocean floor, manipulate objects with remote-controlled tools, and release divers for useful seabed work are changing man's direct contact with the sea. Habitats set up on the ocean floor enable human beings to live and work under water without returning frequently to the surface. The ocean now encompasses more than the surface of the water or the shallows along the shore, for a world under the sea is available for man's use.

The very complexity of ocean uses has increased the potential conflict among competing forces. Modern advances in the sea have produced contests undreamed of when Rome fought Carthage for control of the western Mediterranean. Without breaking into open warfare, fishermen in large factory ships annoy men in primitive boats. New offshore oil installations impede navigation and must be carefully regulated by law to protect ships in traditional sea lanes. New activities may seem innocuous at first because they use space not claimed by anyone else. However, when oil from a leaking well in the Santa Barbara Channel in California, for example, pollutes the beaches and kills wild life, the authorities must acknowledge the conflicts inherent in underwater programs.

The ocean bottom, especially in the North Atlantic, is covered with a network of communication cables, so thick that some lie only a few miles apart. Laying cable is one of the freedoms of the high seas, and maritime nations, since 1884, have accepted regulations for protecting submarine cable. Such installations might seem to be free from interference; yet in 1959, according to a United States diplomatic protest, a Soviet trawler put twelve breaks in five cables lying under a thousand feet of water. The Soviet ship was observed off Newfoundland by an American aircraft, which dropped a note requesting the ship to stop trawling. Later the United States sent an unarmed boarding party of one officer and four men to inspect the trawler's gear. It was unclear whether the cables had been broken accidentally on the bottom or had been pulled up and cut to free Russian nets from entanglement.

The boarding operation was carried out peacefully. Later, diplomatic notes were exchanged between the two governments. The Soviets claimed that American newspaper articles proved that the inspection of the trawler had been undertaken for "provocative purposes," a charge which the United States denied. The affair was settled peacefully after a mild diplomatic storm, and the cables were repaired. However, there is no guarantee that they will not be hauled up or broken by trawls again.

The sea is becoming crowded. Land disputes may be difficult to settle, but nations can at least draw boundaries and enforce claims of sovereignty over specific regions. To avoid chaos on land, some governmental agency must exist in every city and country area. Only in unusual cases, as in the city of Berlin after World War II, have several nations divided jurisdiction over an area. On the sea, boundaries are more difficult to draw. Only in the last hundred years have marine charts become accurate. Nevertheless, pinpointing a location on the open ocean within a one-mile radius remains difficult.

It has always been difficult to enforce law on the sea, or to determine the extent of its authority. Now a new, third dimension must be considered, the space under the surface—the seabed and subsoil of the ocean bottom. And air space above the water adds one more factor to provide possible conflict. Under the circumstances, it is surprising that Law of the Sea is not more confused, and that it can be defined as well as it is.

Who, then, owns the ocean? Who has the right to decide how it will be used, or to adjudicate between conflicting activities in the sea? How have men arrived at a peaceful division of this tremendous area in the past? In an era of rapid change, how can new laws be created to regulate new types of exploitation? After legal order is established on the oceans, what machinery will be available to implement new laws?

Some answers to questions about sea ownership are well defined. Others are nebulous, still evolving under the pressure of events. The State of Florida, for example, definitely owns the seabed of the continental shelf within the three-mile territorial limit along her coasts. In 1958, a diver named Kip Wagner, believing that a Spanish treasure fleet lay in shallow water forty miles south of Cape Kennedy, obtained a salvage lease from the state, by which he might retain 25 percent of everything he recovered. From extensive research he learned that a Spanish fleet loaded with gold and emeralds from South America, coins from the Mexico City mint, and goods from China had been caught in a hurricane in 1715. Of the eleven ships, ten sank, and although the Spanish later returned to the site, they recovered less than half the cargo.

Wagner and his divers brought up three million dollars' worth of gold doubloons and Spanish jewelry, a gold ring with a 2.5-carat diamond, a gold whistle and chain, and fragile Chinese pottery, still packed in clay from its original shipment. In one sale of recovered articles, Wagner collected more than two hundred thousand dollars, his share after the State of Florida had taken three-quarters of the profit.

Treasure recovered from the Spanish fleet off Florida is one example of wealth from the sea, more tangible and exciting than profit from an oil well or an anchoveta harvest but far less important. To obtain wealth from the oceans, whether natural resources or gold doubloons, requires capital investment, something most people will not risk unless their profit is protected. The State of Florida was secure in its ownership before Wagner's divers went down. The United States Congress, after more than a decade of dispute over oil leases, passed laws in 1953 giving the states seabed control within territorial waters and the federal government jurisdiction over the remaining continental shelf.

Traditionally, dominion over the sea was determined by

naked force: a strong fleet could drive off a weaker one. Fishermen, in conflict over the best fishing grounds, were protected by their home countries, which settled such disputes, along with many other questions, through warfare and diplomacy. Although few battles have been fought over ocean rights in the last hundred years, it is unwise to underestimate the role of force—potential or actual—in modern negotiations. Nations which can and will patrol coastal waters have the best chance of enforcing their claims. For example, whenever foreign fishing boats drift into the American twelve-mile fishing zone, the United States immediately has airplanes and Coast Guard ships on the scene to control the situation. Each nation's potential power influences its weight in international councils.

Treaties in which two or more nations agree on compromises, set up regulations, or adjust claims of sovereignty have often followed force in making sea laws. Sometimes these arrangements have been made at the end of wars in which battle decided the victor's right to dictate terms; at other times, friendly nations solved problems before force was necessary.

A more sophisticated method of deciding control of the sea is the regional conference at which delegates from many nations gather to reach agreements regarding individual fisheries or specific ocean areas. Such agreements have had the force of law only for nations that sign them voluntarily, so that their enforcement depends on mutual cooperation.

Now, when the United States and the Japanese disagree about the harvest of king crab in the North Pacific, conflict is avoided by a series of treaties, renegotiated every two or three years to keep a fair balance of rights between the fishermen from the two nations. The Russians and the Japanese—historic rivals over fisheries in the northwest Pacific—make treaties that last a varying number of years and cover every situation that might arise between fishermen of the two countries. A veritable spiderweb of interlocking treaties controls commercial relationships, fishing, oil exploration, and special questions such as the use of channels or narrow straits.

Commissions also are set up for specified areas, such as the northwest Atlantic and the Mediterranean, and for specific fisheries, such as whaling. The sailor who ventures into the open ocean is free from authority so long as he does nothing but sail his boat. The minute he engages in commercial activity, he is limited by as many laws as he would find on land.

The creation first of the League of Nations and then of the United Nations, both based on a belief in cooperation and international law, has led to a number of sea-law conferences attended by many countries. An unproductive conference to solve problems about territorial limits was called by the League at the Hague in 1930. The United Nations sponsored two more at Geneva in 1958 and 1960.

World councils are no longer so concerned about the power of specific nations in one ocean area or another, but have begun to focus on international law. During the last five hundred years, international Law of the Sea has been diffuse, uncodified, and accepted or rejected by nations according to their own needs. While legal authorities defined concepts of law, events on the oceans determined the practical balance of power. Judicial decisions, in national courts and in international tribunals such as the World Court, set up precedents that guided future conduct. Finally in 1958 at Geneva, the United Nations Conference on the Law of the Sea[5] made an important attempt to codify sea law and to discover what precepts the majority of nations would agree to. Such law, constantly modified, gives a fairly fixed response to the question "Who owns the ocean?"

The British, whose mastery of the sea dominated nineteenth-century history, cared little for the opinion of, let us say, Switzerland, Burma, or the Congo, about their actions. It was public opinion at home that concerned them. Africans and Asians were thought to have no political significance. With the present forums of international discussion, however, nations using the sea are driven by a new pressure: world opinion. Force is no longer accepted without question as the basis for determining sea law. Nations with ocean capabilities are expected to feel some responsibility for those lacking them, and although this is not the motive behind many actions, it at least prevents uncontrolled greed in the sea. Nations—including small, emergent, or land-locked countries—expressing their opinions or uniting their forces in an international assembly add a new aspect to the development of the Law of the Sea.

We are in a period of great change in man's relationship to the sea—both in his ability to use it and in his ways of governing it. New ideas filter into discussions about the oceans: that the undersea world is as important as the surface; that ocean resources must be controlled and conserved; that intelligent cooperation in sea use will benefit all nations more than force can possibly help the single country using it; that the international Law of the Sea is as important as comparable law on land; that conflicting uses of the sea must be regulated to the advantage of all; that scientific knowledge is needed to ensure maximum benefit from the oceans. Although some of these precepts are supported more by lip service than by action, they represent goals toward which the world can strive.

There is not, and never can be, a single answer to the question "Who owns the ocean?" History supplies the background, and current events the setting for possible answers. This book relates how the Law of the Sea has developed through the centuries and how it stands today, with emphasis on its changing character. It tells of the technological revolution that has been going on during the past decade: the new vehicles being built to observe and do useful work under water; the new habitats that can be placed on the ocean bottom to allow men to live and work in a deepwater environment; the changing techniques in fishing and fish farming; and the constantly improving drilling rigs that enable men to bring up oil and minerals from the bottom of the deep sea. The new technology is changing so rapidly that it can be described only as it is at one moment in time, for it will be different tomorrow. The law also is changing, although not so rapidly as the technology. Just as law on land has grown up through the ages—being codified occasionally and adapted to new conditions—so the Law of the Sea is developing as a body of accumulated regulations accepted on an international scale. The more men need to use the oceans, the more they will need laws to govern it.

Historical Section

1   Phoenician galley ship

2   Riff pirates attacking a Spanish ship at night

3 Hugo Grotius

4 "Pieces of eight" and jewelry collected from sunken ships off Florida coast

5 Calico Jack

6 Edward P. Teach, known as "Blackbeard"

7 Federal boat ramming a rum runner

8 U. S. Customs agents unloading whiskey from fishing trawler captured by U. S. Coast Guard

9 Intelligence-gathering ship *U.S.S. Pueblo*, captured by North Korea

10 Memorial to Captain Cook in Honolulu

11 Lieutenant Matthew Fontaine Maury

12 Afro-Asiatic group holding informal meeting at 1958 Geneva Conference on the Law of the Sea

CHAPTER 2

The Historic Sea

The recorded history of man's use of the sea goes back to a time, almost three thousand years before Christ, when the Egyptians began the first shipbuilding industry, using timber brought by raft from Lebanon and Syria. In spite of their shipbuilding, the Egyptians never became great navigators on the Mediterranean, although, like the Sumerians before them, they made full use of the river that dominated their country. They exerted their power chiefly on land and considered sea trade of secondary importance. Away from the Nile, they were never a great fishing nation. Their priests and nobles scorned fish, a common food among the poor, probably because it came from the Nile and tasted of mud.

The greatest seafaring people of the Ancient World were the Phoenicians, who traded with countries as far distant as India and the northwest coast of Spain. Because their homeland on the eastern shore of the Mediterranean was small, they established colonies, some of which became more important than the Phoenicians' own land. Carthage, a city they founded on the coast of Africa, dominated all the trade routes of the western Mediterranean until it came into conflict with Rome's rising power.

One Phoenician colony on the island of Rhodes is the place where the earliest rules for ships are supposed to have developed. Although the existing versions date from a later period, Rhodian law apparently was used by Greek and Roman sailors before it was incorporated into Byzantine law. Its early regulations were probably very simple. If the mast of a ship had to be heaved over the side during a storm to save the vessel from sinking, everyone aboard, including merchants traveling with their cargoes, had to help finance a new mast. Other rules designated punishment for one sailor's striking another and stipulated when sailors might safely leave the ship to sleep on shore. Such rules were not established by lawyers but grew up informally over a long time and were exchanged among seamen of different countries.

The importance of Rhodian law lies in its international character. In the Ancient World, laws on land varied tremendously from one city-state to another, but sea regulations were disseminated widely and were accepted aboard the ships of many nations. The Law of the Sea—whether concerned with merchant shipping or with sovereignty claims—has always been international in character because of the universal nature of the sea itself.

Although the Phoenicians were famous more for their navigation than for their fishing, fish must have been important to them. Their small country had two major cities, Sidon and Tyre. The Phoenician word "Sidon" means fish, and Tyre was named after the man, Tyrus, who was the traditional inventor of fishing tackle. Although they were never a great military power on land, the Phoenicians supposedly developed the first armed navy to protect their distant trading vessels.

An early king of Crete, called Minos (the name was probably a title, like "pharoah" or "tsar," rather than a personal designation), is credited with having created the first regular navy and with having claimed sovereignty over the sea surrounding Crete. His claim rested partly on his reputation for keeping the waters free of pirates. In later history, many nations based their ownership of the sea on their ability to patrol certain areas, helping ships in distress and protecting them from pirates. On the whole, the later Greeks had little say about laws governing the sea. They understood sea power—the capacity of one fleet to overcome another at a given time and place—but their city-states seldom tried to dominate the sea. Fishing produced an important part of their diet, and they knew how to preserve and use salt fish. In both Greece and Rome, there were periods when certain fish were considered great delicacies. Nobles and rich merchants competed with each other, trying to serve the fanciest and most expensive seafood.

The great lawyers of Rome were concerned with the kind of law that governed the sea and with the theory of ownership that applied to it. They studied all kinds of ownership and considered the problem of classifying the sea and the seashore. In the *Institutes of Justinian* (533 A.D.), a summary of Roman law, one section says:

For some things are by natural law common to all, some are public, some belong to a society or corporation, and some belong to no one. But most things belong to individuals. . . . Thus, the following things are by natural law common to all—the air, running water, the sea, and consequently the seashore.[1]

Two categories of things listed by the Romans are usually connected with the sea: things that belong to everyone (*res communis*) and things that belong to no one (*res nullius*). Modern lawyers still use these two classifications in describing the sea and argue about which description applies at the present time. Many writers say that the category of things belonging to no one is not in keeping with the freedom of the seas because, at some time in the future, people may claim what no one owns now. The category of things belonging to everyone is considered a more positive basis for the theory of freedom of the seas.

Whatever Roman lawyers may have had in mind, it is fairly clear that in ancient times the sea was considered an area of freedom beyond the laws of individual city-states. The Romans upheld this theory in their laws; in practice, they used the sea to their own advantage. When they came into conflict with Carthage over control of the western Mediterranean, they developed a strong navy and defeated the Phoenician colony in a series of wars. After that time, they controlled trade, traveled by water, moved troops and supplies by sea, and expanded their empire as far north as England.

The Mediterranean, in early Roman times, was harassed by pirates, who, by themselves, were almost strong enough to form a military power. Julius Caesar, as a young man, was captured by pirates and held for ransom. Waiting for the money to arrive, he told his captors frankly that, if he were freed, he would capture them and have them crucified. When his ransom was paid, he immediately carried out his threat. When the pirate raids became so numerous that they threatened Rome's grain supply, the famous General Pompey finally drove the marauders from the sea. To accomplish this, he divided the Mediterranean into districts, conquered the outlaws in one area, appointed an official to keep the peace there, and then moved his ships to a new part of the sea. The Romans often called the Mediterranean "our sea" (*mare nostrum*), meaning that they controlled it, rather than that they owned it.

With their precise legal minds, the Romans classified fish as animals wild by nature (*animales ferae naturae*), in the same category as bees and deer. This meant that such creatures could be owned only when they were actually in someone's possession. If a man had bees in his hive, he owned them. As soon as they flew away, they reverted to their natural wild state. A man owned fish only when they were in his net. Fish still swimming in the sea were wild animals, not subject to ownership.

Roman ships traded throughout the Mediterranean, and the navy could carry Roman soldiers to the farthest parts of the known world. But Roman power depended chiefly on the armies and on the network of overland roads built to control the empire. When the Romans' power declined, the Byzantines, and later the Moslems, controlled the Mediterranean. Trade decreased until, in the early Middle Ages, the countries of northern Europe had little access to Africa and the Near East. The English Channel and the Baltic Sea continued to be avenues of travel, but even journeys as short as that from England to France were dangerous in the ships of the day.

Feudal law in the Middle Ages rested squarely on ownership of land, for all wealth and power depended on grants of property handed down from one authority to another. As early as Anglo-Saxon days, some English kings styled themselves rulers of the sea, and Edward III, in the fourteenth century, expected foreign ships to salute his vessels because he was "King of the Seas." Because naval warfare was so uncertain at the time, no one could prove or disprove the claim. Fishing rights near the coast were usually considered part of the king's property, which he could grant to others and on which he could collect taxes. Whales and other animals washed ashore in England belonged to the king and queen.

One concept which developed slowly during the Middle Ages was that the sea near the coast of a country belonged to that country. This idea of ownership of adjacent waters remained vague. No effort was made to define how wide a strip of water came under the definition. But it was apparent from an early date that a coastal nation had some control over what took place in the sea immediately along its shore.

In the great age of discovery, during the fifteenth and sixteenth centuries, many nations claimed ownership of parts of the ocean. The Crusades had established new contacts between northern Europe and the Mediterranean world. Improved ships could undertake long voyages. Soon after the Portuguese sailed down the coast of Africa, Columbus opened up the Atlantic. The nations that took possession of new lands also assumed that they had jurisdiction over the seas they crossed.

In the Mediterranean area, Venice held the Adriatic Sea, and Pisa and Genoa maintained ownership over waters near their coasts. With parts of the Mediterranean in the hands of the Turks at Constantinople, there was little freedom of the seas in that region until after the sixteenth century. In the Baltic and North Atlantic, Scandinavian countries claimed parts of the sea as far as Greenland. Spain and Portugal, the two great nations of navigators, wanted complete control over waters in their areas of discovery. Before the end of the fifteenth century, a number of papal bulls upheld Spanish and Portuguese claims to the sea, and in 1494 a treaty drew a line dividing the oceans between the two countries. Portugal gained control of the waters off Africa and in the Indian Ocean, while Spain's sphere included most of the Americas.

It was soon apparent that no papal bull could control the use of such vast areas of the ocean. English sailors raided Spanish shipping and were rewarded by Queen Elizabeth I when they returned home with captured treasure. Although England did not give up her rights to sovereignty in such areas as the English Channel, Elizabeth said that no one could stop her sailors from using the ocean because "the use of the sea and air is common to all; neither can a title to the ocean belong to any people or private persons, forasmuch as neither nature nor public use and custom permit any possession thereof."[2]

The first writer to stress the concept of freedom of the sea was a Dutch lawyer, Hugo Grotius. In 1609, Holland was disputing Portugal's claims to the Indian Ocean. Grotius, defending his country's use of the sea, wrote a Latin treatise on the subject. He said that the sea cannot be owned, first, because no one can take actual possession of it, and second, because it is inexhaustible. On land, a territory can be surrounded by soldiers, possessed, and occupied; on the sea, ships cannot occupy or hold any part of the sea except for the short time they sail on it. The fish of the sea, Grotius said, are inexhaustible. Regardless of how many fish men remove from the ocean, endless numbers will still be left for others. Accordingly, the seas cannot be owned; they are open to everyone's use.

In 1618, an Englishman, John Selden, disagreeing with Grotius's concepts, wrote in favor of the closed sea. He argued that nations, especially England, could and should own part of the sea. When he wrote that England held sovereignty over the waters surrounding the British Isles, his position was a difficult one, for his country seemed to want to have its cake and eat it too—both to possess the sea near home and to have complete freedom elsewhere. Many writers of that day realized that some arrangement had to be made that would allow countries to control the waters along their coasts. Simultaneously, there was strong support for Grotius's doctrine of freedom of the seas. Men were groping for legal principles that would combine the best points of the two ways of regarding the sea.

Gradually, the idea developed that the waters along its shore, out to some accepted distance, were to be considered part of a nation's territory. Writers discussed how wide these territorial waters should be and what kind of jurisdiction the coastal nation had over them. A Dutch lawyer, Cornelis van Bynkershoek, popularized the idea that nations could possess the water off their coasts as far as a cannon could shoot. Although claims to the sea along the shore partially reflected a desire to control commerce and fishing, national safety was also a major reason why countries wanted to keep foreign ships at a distance from their beaches. The cannon-shot rule was a practical way of saying that a nation could control the sea as far as its shore batteries would reach; and, at a time when charts were inaccurate, it was an easy rule to apply. If a ship was close enough to be struck by a cannonball, it was too close. In the same era, many nations gave up any attempt to control large areas of the sea, reducing their jurisdiction to one sea league from shore. In the Scandinavian countries a sea league was four miles, and the territorial limit in that part of the world has generally remained at four miles. In most other nations, including England, the sea league was three miles, which coincides very closely with the distance most cannons could shoot. The United States first claimed a three-mile territorial sea in 1793. The two ideas—the cannon-shot rule and the sea league of three miles—gradually merged, and the concept of territorial waters became widely accepted.

In the seventeenth century, while legal minds weighed basic principles, England maintained that she had sovereignty over the sea surrounding her shores and still insisted that vessels of other nations dip their flags to her ships in recognition of this status.

But the Elizabethan age of Drake and Hawkins was over. Some men were fighting for religious beliefs; others pursued their interest in profits from fishing and commerce. The Dutch, in their extensive trade with the Far East, had become a major maritime power. Their ships carried most of the goods of Europe, and their seamen developed the profitable herring fisheries of the North Sea. Tradition has it that Amsterdam is built on herring bones. The Dutch herring fleets introduced the use of trawl nets, bringing them close to the shores of England and Scotland—a practice that infuriated local fishermen.

In 1651, England passed a law requiring that goods brought

into her ports be carried in English vessels or in those of the country of origin. Dutch middlemen could no longer carry on their trade. In addition the Dutch were expected to get an English license to fish for herring. These restrictions precipitated the first of three maritime wars between the two nations. The underlying cause was commercial rivalry, but the immediate dispute began over the Dutch refusal to dip their flag in acknowledgment of English sovereignty on the water. Tradition says that, after one victorious engagement, the Dutch admiral nailed a broom to his mast to show the world that he had swept the English from the sea. Nevertheless, the English were often successful, and years of indecisive naval battles exhausted the Dutch.

Although the Anglo-Dutch wars of the seventeenth century settled very little, they had some side effects. The colony of New Amsterdam passed from the hands of the Dutch in one treaty and became English, as New York. After the English and Dutch stopped fighting each other, England, turning to expansion in more distant areas of the world, lost interest in claims to exclusive use of the waters near her coast. In time she became the great champion of narrow territorial limits and freedom of the high seas.

During the seventeenth and eighteenth centuries, two types of raiders sailed the Atlantic: privateers, who carried commissions from their governments to prey on enemy ships; and pirates, who attacked other vessels indiscriminately. The distinction between the two was not always clear, for both Francis Drake and John Paul Jones were called pirates by the nations whose ships they captured. Pirates had no nationality, but needed help from people on shore to sell their stolen goods and to refit their ships. Edward Teach, the famous Blackbeard, for example, had his winter quarters on the Carolina coast and shared his plunder with colonial officials. Since they preyed on ships of all nations, pirates could be captured by men of any country and tried in any port. Usually, however, it was recognized that they deserved a fair trial, despite their lack of citizenship.

Because pirates put themselves outside the law of any nation, men soon realized that each peaceful ship should be subject to the laws of one nation and should carry that nation's flag. Under present international law, every ship must be registered in a country, carry its flag, and be subject to its regulations for life at sea. When a small pleasure craft or a great ocean liner raises an ensign, the flag is not for decoration. It signifies that the vessel is under national jurisdiction and hence within the law.

Piracy in the Atlantic thrived on unsettled conditions. When it finally died out, it did so not only because the British and American navies captured the more notorious pirates, but also because settled conditions discouraged men from entering a career no longer safe or profitable.

Until 1830, piracy continued in the Mediterranean, where the Barbary pirates set up kingdoms in North Africa and raided the ships of all non-Moslem nations. They were suppressed by the efforts of many countries—the United States, England, Holland, and finally France, which took possession of Algeria, destroying a major center of pirate trade. In Asia, pirates were active in the China Sea until the early years of the twentieth century; as late as 1951, indeed, pirates in the China Sea tore up and made off with three and a half miles of the Danish telephone cable that ran between Hong Kong and Amoy.

Maritime problems influenced relations between England and the United States both before and after the Revolutionary War. Beginning in 1660, Parliament passed a series of laws called the Navigation Acts that strictly controlled the trade of the colonies, forcing Americans to ship their goods to England before they could be re-exported to other parts of the world. The laws were not strictly enforced and did little to hamper trade. But after the Treaty of Paris (1763) ended the conflict between England and France in North America, Parliament attempted to control sea trade in the colonies and to raise revenue to pay for troops used in North America. The Sugar Act of 1764 was intended to affect traffic in lumber, food, molasses, and rum between New England and the West Indies. Opposition to such control of maritime activity reinforced colonial objections to the Stamp Act and the tax on tea, and helped to cause the War for Independence.

After 1800, Britain and America again came into conflict over freedom of the seas. The United States protested against the impressment of American seamen when British ships, stopping vessels on the high seas to search for sailors that had deserted from the British Navy, seized United States citizens as well. It also complained that British ships were entering United States territorial waters without permission.

To defend its rights on the sea, the United States fought the War of 1812, pitting its new, untried navy against the widely acclaimed superiority of the British fleet, and winning a number of victories. Peace came in 1814 when the two nations realized that the war harmed both sides and benefited no one. In practice, the British stopped impressing American seamen, although, in theory, English ships still claimed the right to stop and search vessels on the high seas.

During the first half of the nineteenth century, both England and the United States patrolled the seas to stop illegal traffic in slaves. Over two million black people had been imported to the British colonies and the West Indies in the hundred years before the Revolution. The United States outlawed the importation of slaves in 1808, and England, which had already passed similar laws, gradually persuaded most of the European nations to follow her lead. Groups of ships, such as the United States African Squadron, tried to enforce the law by searching suspected vessels. However, so long as slavery remained legal in many parts of the world, slave traffic was difficult to combat. Efforts at control made conditions worse for the slaves, who were crowded into overloaded ships and thrown overboard when the vessels were threatened by the authorities. It is said that three times as many black people as before were exported from Africa, and of these, two-thirds were murdered on the high seas.

By 1850, when slavery had become illegal in most European colonies, the African Squadron had become more active. Although slavery was allowed in the United States until after the Civil War, and in Portuguese possessions until 1878, by that time traffic in human cargo had died out. In modern international law, provisions specifically outlaw slave trade on the high seas.

In addition to controlling piracy and the slave trade, maritime nations tried to curtail coastal smuggling by stopping vessels on the high seas. In England, smuggling laws date back to 1351, when it became treason to bring counterfeit money secretly into the country. In the North American colonies, men smuggled liquor illegally to the Indians, carrying bottles in their boots for sale; hence the use of the word "bootlegger" for anyone who transports liquor against the law.

During Prohibition in the United States, liquor was

brought from abroad in ships that waited just outside the territorial limit, while small, fast boats, many of them armored, carried the cargo to shore. Over three hundred rum-runners were active in the trade in 1924. To control smuggling outside of the territorial limit, the United States made a series of treaties with other nations that allowed the government to stop ships at sea a certain distance from shore. Thus the control of smuggling became an exception to the theory of absolute freedom on the high seas. It has become customary for countries to extend their limited jurisdiction, sometimes for a specified time, beyond their territorial sea to control smuggling. Jurisdiction is also extended to stop illegal immigration and to enforce regulations against pollution.

After the final defeat of Napoleon in 1815, no power was strong enough to challenge the British Navy. Captain Cook had explored and charted the farthest reaches of the Pacific. Admiral Nelson had won victories at Aboukir, Trafalgar, and Copenhagen. English ships traded around the world, and British laws reducing tariffs tended to encourage the trade of other nations. After steamships became common in the 1850's, England had scattered bases supplied with coal to serve as refueling stations. England, followed by France and Germany, used sea power to open and develop colonies, which then became additional markets and sources of raw material. Ships with big guns, anchored in foreign harbors, were powerful forces in international policy.

During the nineteenth century, the great maritime powers favored narrow territorial waters and complete freedom of the high seas. They wanted straits and passageways kept open to navigation and felt that coastal nations, if they claimed more than a three-mile marginal strip, would be able to close off gateways like the Strait of Gibraltar, which is eight miles wide. (Four miles of territorial sea on each side would eliminate the strait as part of the high seas.)

Therefore, England and the other maritime nations brought pressure on coastal countries to open passageways to international shipping. Denmark, which traditionally had collected money from vessels entering the Baltic, opened her straits after various nations paid a lump sum as a substitute for future dues. At the tip of South America, the Strait of Magellan caused conflict between Chile and Argentina until 1881, when the two nations signed a treaty guaranteeing freedom of navigation to all ships. Two of the most important straits in the nineteenth century were the Dardanelles and the Bosporus, between the Mediterranean and the Black Sea. Turkey claimed the right to close the straits to all vessels, but Russia compelled her, in 1774, to open them for merchant ships. Warships were denied entrance until after World War I, when a long period of negotiation on the subject began. The suggestion that the Dardanelles be declared an international area has never been accepted by the nations involved.

The great maritime nations, during the nineteenth century, were concerned not only with curbing abuses on the sea but also with using the ocean as a pathway to knowledge. From 1831 to 1836, the English ship *Beagle,* on a surveying expedition, touched at many points in South America, the Galápagos Islands, and islands of the South Pacific. The ship's naturalist was a young man named Charles Darwin, who developed, from what he learned on that voyage, ideas that were to shake the world. The United States sent out the Exploring Expedition of 1838–42, led by Navy Lieutenant Charles Wilkes, which skirted Antarctica, visited the Society and Fiji Islands, and surveyed what became the west coast of the United States. Scientists aboard the six vessels brought back information that greatly increased man's knowledge of the Pacific.

Mapping and charting the oceans became a major activity during the nineteenth century. When Columbus sailed, marine maps were carefully guarded because they contained valuable trade secrets. Now, however, England took the lead in charting the seas and making the information available to all. The Survey Service and Hydrographic Department of the British Admiralty, founded as far back as the 1750's, is still in existence.

An American naval officer, Matthew Fontaine Maury, one of the first oceanographers, became world-famous for his charting of the seas. After giving up sea duty because of a lame leg, Maury was assigned to the Navy Depot of Charts and Instruments, where he gathered logbooks from ship captains around the world and used the information to chart wind, weather, currents, and temperatures in the ocean. Maury's charts helped to reduce the sailing time from New York to San Francisco from 180 days to 133 days. (In 1855 he published *The Physical Geography of the Sea,* which is considered the first text book in oceanography.)

After thirty years' service in the Navy, Maury was about to be retired with the low rank of lieutenant, but friends and well-wishers finally convinced top naval officers that the man in the U. S. Naval Observatory and Hydrographic Office (the new name for the Depot of Charts and Instruments) was a famous scientist. Therefore, before he resigned to join the Confederacy during the Civil War, he received the rank of commander. Maury spent the last years of his life as Professor of Meteorology at the Virginia Military Institute, and died in 1873.

After the introduction of steamships, traffic control became necessary in crowded sea lanes, and a body of international maritime law was developed to govern "rules of the road." The International Rules for Prevention of Collision at Sea, adopted internationally in 1889, were based on British rules of 1862. They are accepted by all maritime courts as the standard by which to judge who is at fault in a collision. The rules explain how ships must navigate with respect to each other, which lights must be shown, and what signals given. It is ironic that, although the British originated the rules, ships at sea keep to the right. On English roads, traveling on the left is a custom that goes back into forgotten history. Some say it originated because men on horseback wanted their sword arms to be on the same side as an approaching horseman. Nevertheless, in most nations of the world except Britain, automobiles are driven on the right side. The British, however, created rules of the road at sea in accord with the prevalent custom.

Other international regulations have been created by maritime conferences and, revised quite frequently, continue to govern ships at sea. After the sinking of the *Titanic* in 1912, public demand for greater safety in ocean travel resulted in a series of International Conventions for the Safety of Life at Sea (SOLAS). The latest version of its regulations, adopted in 1960 (SOLAS 60), provides for safe construction of ships, watertight bulkheads, life-saving appliances, fire-fighting equipment, radio and direction finders, and regular emergency drills. It has been modified by improved Fire Standards for Passenger Vessels (1966). These have not been approved internationally, but are required for all ships sailing from United States ports with United States nationals aboard. The International Load Line Convention specifies how cargo ships are to be loaded. Modern vessels have a mark on the hull amidships—a circle with a line through it, called the Plimsoll mark; if the waterline rises above this point, the ship is loaded too heavily. Loading regulations have saved crews from sailing on "coffin ships"—freighters or tankers so heavy that they break in two or sink. In a world in which the great powers rarely seem to agree, the absolute necessity for safety on the water has forced maritime nations to accept international rules to protect everyone.

Agreement is also being reached by the world's nations on many important rules of international Law of the Sea. The United Nations Geneva Conference in 1958 established four conventions or agreements that have since been ratified by a number of countries. The Convention on the High Seas defines four major freedoms of the sea: the right to navigation and fishing, the right to lay cable and pipeline, and the right to fly over the water. For many years ship owners, to avoid heavy taxation or strict safety rules, have registered their vessels in small nations such as Liberia or Panama, and flown what are known as *flags of convenience*. The Geneva Conference declared that a genuine link must exist between a nation and ships that fly its flag. The High Seas Convention also set up rules concerning slavery, piracy, the control of smuggling, and disposal of radioactive waste.

In a convention on territorial waters, the Conference established some of the rights of coastal nations inside their territorial limit, and defined *innocent passage* as the right of a foreign ship to navigate in territorial waters provided it does nothing prejudicial to the peace, good order, or security of the coastal state. Fishing boats in innocent passage are required to obey coastal regulations, which usually prohibit fishing. Submarines must navigate on the surface and fly their national flag.

The 1958 Geneva Conference failed to reach agreement on a definite width for territorial waters, just as the Hague Conference had failed in 1930. The United States would like to see this width defined by international law, but the Soviet Union believes that each nation should have the right to determine its own territorial limit. The second Geneva Conference in 1960, called specifically to seek agreement about territorial width, explored various compromises, such as a six-mile territorial zone with a further six miles of fishing rights, but it failed to accomplish its goal. Consequently the United States, Britain, and many other countries continue to claim three miles, the Scandinavian nations claim four, the Soviet Union and most of its followers claim twelve miles, and some South American nations claim two hundred miles.

In 1970, the U.S. State Department announced that it would favor some future agreement on a twelve-mile territorial zone, provided that straits necessary to international navigation were kept open.

Some narrow straits necessary to navigation fall within territorial waters. Here the question of innocent passage becomes of primary importance. Currently, one of the most disputed areas in the world is the Strait of Tiran, the passage between the Gulf of Aqaba and the Red Sea. Egypt considers this part of her territory and has controlled it from Sharm El Sheik, a desert post on the Sinai Peninsula. Because Israel's port of Elath on the Gulf of Aqaba is her only outlet to the Red Sea and the Indian Ocean, Egypt has been able to harass Israel by closing the Strait of Tiran. When Egypt did this in 1957, Israeli forces moved into Sinai and drove the Egyptians out of Sharm El Sheik. Israel agreed to leave Sinai only because the presence of United Nations troops guaranteed that the strait would remain open. Then, in 1967, Egypt ordered the United Nations forces out of Sinai and again closed the mouth of the Gulf of Aqaba. This was a major cause of the Six Days' War, in which Israel drove the Egyptians from Sinai a second time. Most of the world realizes that the Strait of Tiran remains a source of acute danger, but efforts to find a permanent solution have thus far been unsuccessful.

When nations claim to own or have authority over part of the sea, what is the basis for their claims? Some assertions are realistic, others are without foundation. Most fall into several categories:

Protection of shipping. Some rulers, beginning in antiquity, have based their claims to sovereignty over the sea on the fact that they patrolled the sea, rescued ships in distress, and put down pirates. When true, the claim was often acceptable to other nations.

Adjacent waters. Nations have always tended to claim parts of the ocean lying along their territory. Claims to coastal waters assume that the sea is an extension of the land.

National safety. Marginal strips of water naturally protect the land they border. When a cannon shot reached three miles, nations wanted foreign ships to keep that distance from the coast. Today, with modern weapons that can be projected across hundreds of miles, most nations still want a safe buffer zone of water along their shore.

Historic usage. When fishermen have used a certain bay

for centuries and their right to do so has never been questioned, they come to consider the bay their special territory. As in many other kinds of jurisdiction, long usage invests an action with the force of law.

Conservation of resources. As it becomes apparent that ocean resources are not inexhaustible, certain governments claim jurisdiction over parts of the sea in order to conserve fisheries. Conservation claims are supposed to rest on sound scientific information, but sometimes this is questionable.

Exploitation of resources. Men have discovered resources in the sea, such as oil, that must be developed to be used. No oil company would spend money on an offshore well unless some authority guaranteed ownership of the seabed and prevented foreign companies from putting down wells nearby. Hence some claims to jurisdiction rest on the need for national authority if resources are to be exploited.

Economic necessity. A very recent basis for claims is a country's need for the sea. If a modern nation depends on fish for food, if it counts ocean resources part of its wealth, if the sea supplies jobs for its men, then such a country may demand special rights to the use of the sea.

Whatever reason nations give for their claims to the sea, there is continuing conflict between forces that want to bring great areas of the ocean under national domination and forces that try to preserve the freedom of the seas. As increased wealth comes from the sea, the influence of the national-sovereignty principle tends to increase. But the need for free navigation, both in the water and in the air above it, causes nations to advocate unrestricted use of the ocean. The Law of the Sea reflects the constant need to retain a balance between the two.

CHAPTER 3

The Three-Dimensional Sea

The great oceans cover vast parts of the earth; four-fifths of the southern hemisphere and three-fifths of the northern one are made up of water. If we ignore the shallow, lighted areas along the continental borders, we can say that the deep oceans occupy half the surface of the globe. Man lives and finds his food in a rather small area of the earth, often unaware of what happens in that other world, the sea.

The ocean bottom is divided into three zones: the continental shelf along the shore; the continental slope, where the seabed drops off sharply; and the deep ocean floor. How to define the exact limits of the continental shelf is an important point in the Law of the Sea, because claims to oil and other minerals in the seabed depend on legal ownership of this area. The 1958 Geneva Conference reached agreement on a three-fold definition. The shelf is adjacent to the shore

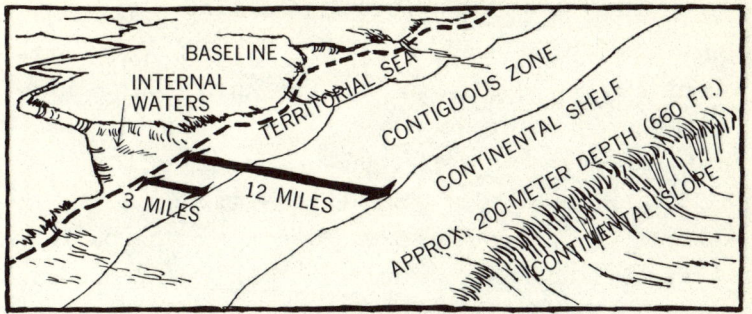

DIAGRAM OF CONTINENTAL SHELF

and reaches to a depth of 200 meters, or as deep as the natural resources can be exploited.

The width of the continental shelf varies around the world. On the east coast of the United States, north of Cape Hatteras, it extends approximately 150 miles out to sea. On the Pacific coast of South America, the land drops off steeply and the shelf is very narrow. It is widest along the Arctic coast: in the Barents Sea, it is 750 miles wide.

The shelf is covered by shallow waters, into which the sun penetrates, making a favorable environment for plant and animal life. Sediment coming from the land covers much of the shelf floor. At one time, parts of the shelf were above sea level. Fishermen, trawling the Dogger Bank in the North Sea, have brought up bones of land animals and bits of trees, which scientists believe were buried long ago by rising water. In places, during the Ice Age, the shelf was carved by glaciers into underwater hills and valleys. Rivers pouring into the sea have left their mark on the shelf.

Where the continental shelf drops off steeply, the continental slope begins. Here the bottom is made of bare rock, clay, mud, and sand. Plant life has disappeared because sunlight does not penetrate so deep. Fish survive on the nutrients that drift down from above. The continental slopes are the walls of the deep sea, rising in some places as high as 30,000 feet above their bases. Deep V-shaped canyons cut into the slope. Some, like the Hudson Canyon off New York, are thought to have been produced by rivers flowing into the sea; others may have been produced by heavy mud flows at a time when the sea level was low and glaciers were advancing.

The deep sea floor—the abyss beyond the slopes—is a varied terrain, with plains, steps, mountain ranges, seamounts, depressions, and trenches. The Atlantic Ridge, running from Iceland to the southern part of Africa, is only one of several mountain chains in the sea. Where there are arcs of islands in the Pacific, there are also deep trenches: the Philippine Trench, the Mariana Trench, the Bonin Trench —all canyons dropping below the level of the sea floor. The seabed of the Arctic Ocean has not been charted extensively, but it, too, contains mountains and ridges, canyons and plains. In many places throughout the oceans, seamounts rise to varying depths beneath the ocean surface.

The water of the sea is moved by tides and currents. Tides respond to the gravitational forces of the sun and moon; currents are rivers in the sea, controlled by the spinning of the earth, by surface winds that blow regularly in one direction, and by the influence of both sun and moon. Near the equator, the sun heats the ocean surface, producing expanded, heated water that moves toward the poles, while heavier Arctic water moves back toward the tropics along the sea bed. The trade winds driving westward across the oceans produce currents that, because of the earth's rotation, move clockwise in the northern hemisphere and counterclockwise below the equator. The Equatorial Current in the northern Atlantic turns right near the West Indies and moves north as the Gulf Stream, carrying warmth to the eastern United States and Europe. In the South Atlantic, the Brazil Current moves down the coast of South America until it meets the Benguela Current coming from Antarctica. Pacific currents follow the same pattern, clockwise north of the equator, counterclockwise south of it. One of the most important, the Humboldt Current, carries cold water north along the west coast of South America. Here it brings up rich minerals from the ocean bottom and produces favorable conditions for thriving sea life. In turn, the Kuroshio Current pushes warm water from the Pacific equator northwest and produces a favorable climate for the Philippines and Japan.

In past centuries, men stayed primarily on the surface of the ocean, mapping and charting it in two dimensions. If they claimed any part of the sea, they possessed it by measuring off a flat area—so many miles in one direction, so many in another. Mariners were concerned with ocean depths only because reefs and shallows were dangerous; the bottom was a place to be avoided, not a world to explore. In a few places, men took riches from the seabed—in the sponge fisheries off Tunisia and the pearl beds near the Arabian coast, where diving had been taking place for centuries. But they remained close to shore and were limited by the short time a man could hold his breath. Early laws or agreements were made about waters along a coast or about navigation rights across an ocean. When men began to go under the surface of the water, their concept of the sea and the laws governing it began to change.

Jules Verne envisioned an underwater world long before modern diving techniques were perfected. In the 1870's he created the romantic, diabolical Captain Nemo, who ruled an establishment on the ocean floor. It had a habitat with picture windows, an underwater garden, traveling vehicles, lock-out chambers, and divers who could move freely about in the water. The first scientific steps had been taken that would make some of these things possible, but for people of that day they were all fantastic dreams.

Schemes for diving apparatus and underwater vehicles go back a long way in history. Many early attempts were severely limited by the materials and techniques available at the time. When Edmund Halley, the British astronomer, tried to design a diving apparatus in 1716, he could not use rubber hose, something we take for granted. His diver was expected to use a hose made of leather, sewed together and waterproofed with a mixture of oil and beeswax. By 1830 it was possible to make the first practical diving gear. Developed by Augustus Siebe, it used a closed suit with a metal helmet, an air-intake hose and a regulated outlet valve. Although the suit was improved over the years, the principal ideas employed did not change greatly. During the second half of the nineteenth century, helmeted divers were used in many places, chiefly for salvage work. A ship carrying gold, sunk in shallow water, was the dream of helmeted divers, and such things occasionally happened.

Laws for helmeted divers were the standard rules of salvage, developed over a long period of time. Anyone who saved a ship or cargo at sea, except the professional members of the Coast Guard, was entitled to a share of the goods recovered. This was never set at a fixed percentage, but was decided in each case according to the value of the property recovered and the risk or effort involved in the salvage. Before working on a sunken wreck, a diving company made an agreement with the owners about the division of any property retrieved. An ancient wreck was considered the property of the person who reached it first.

There were definite physical limits to what a helmeted diver could do. He could not go beyond a certain depth and he depended on compressed air pumped from the surface. Safety standards had to be worked out about how far down he could go, how long he could stay, and how slowly he must return to the surface. There were even greater limitations on how a diver could move about, for he was at all times connected to the mother ship. He could move around on the bottom only as far as the length of his air hose allowed. In his weighted, clumsy equipment, he moved primarily up and down. The helmeted diver could do useful work under water, but was not capable of wide, complex movement in three dimensions.

In the first half of the twentieth century, new scientific and technological developments allowed men to use the world beneath the ocean's surface. In 1900, John Holland, the father of the United States submarine, delivered his first acceptable submarine to the United States Navy; a half century later, in 1955, the Navy launched the first atomic submarine. The tremendous strides made in underwater technology during those fifty years have changed man's thinking about the sea. Some new inventions were tentative at first, and their future use was unclear, but most advancements answered practical needs. Two world wars spurred new designs and new engineering. In each case, great advances were made and techniques perfected in a few short years. Between the two wars, in the twenties and thirties, public interest in science and exploration stimulated new inventions.

The beginnings of the conventional military submarine go far back in history. A semi-submerged vehicle was used during the Revolutionary War but did not succeed in causing any damage to the enemy. Robert Fulton designed a submarine, which he offered to both the French and the British governments at various times, but it was never accepted. British admirals were horrified by any invention that might destroy existing means of naval warfare. In the Civil War, a submarine was used by the Confederate Navy in an attempt to break the North's blockade. It succeeded in blowing up one Union ship, but had a history of disaster, drowning a number of men.

John Holland was responsible for the idea of using water ballast as a means of submerging, and he invented horizontal fins to control the angle of dive. In the first submarine that he built for the Navy, he was hindered by official insistence that the boat be propelled by steam while on the surface. Holland realized that this would never work. Before each dive, the crew would have to extinguish the fire in the boiler and either wait until it cooled or else dive and be roasted by the heat. Holland finally delivered a successful ship to the Navy, and the Electric Boat Company, which he founded, is still in business making submarines for the government.

In Holland's day, another man, Simon Lake, made major advances in submarine work. Wanting to produce submarines for peaceful uses, Lake designed and built one that could move along the bottom and allow divers to go out into the water. This boat was used in commercial salvage, but nothing more is heard of its development—perhaps because the cost could not be supported by a private individual. Submarines were mainly used in warfare, becoming highly efficient weapons during World War I. By World War II, they were completely developed fighting machines. Nations had the money, and the war supplied the incentive to produce tremendous advancements. As technology became highly refined, submarines became expensive. They were designed to do nothing but fight. Usually they were built with government money for government purposes.

The conventional submarine is limited in its ability to explore or use the space under water. It cannot descend to great depths. Except as a last resort in times of danger, submarine skippers avoid the bottom, which is full of hazards—outcroppings, irregularities, soft mud, and sudden canyons that can damage an underwater vessel. The men inside a submarine have almost no contact with the water around them, for there are no portholes. Although present-day submarines carry instruments that enable them to detect activity going on outside, their main purpose is still to fight the enemy and escape detection, rather than to explore or work on the bottom.

In the late 1920's and early 1930's, William Beebe became interested in going deeper into the ocean than existing submarines could go, to observe what was happening in the deep sea. Beebe had done a great deal of shallow diving in a helmeted suit with an air hose and had studied underwater marine life in several parts of the world. Fortunately, he met

Otis Barton, an engineer, who had developed a spherical cabin that could be lowered into deep water on a steel cable. Beebe christened this machine the bathysphere.

The sphere Beebe used was only four feet nine inches in diameter—a space that would hold two men in a cramped position. Much care was used in making the steel cable and in designing the fitting that attached the cable to the sphere. The entrance, barely fourteen inches wide, could accommodate only a small man; the door that closed it, weighing four hundred pounds, had to be lifted with block and tackle. The greatest point of danger was the opening in the sphere where telephone and other electrical cables entered. On one trial with the sphere empty, this opening failed to remain sealed, and the sphere was hauled up full of water under pressure. When the door was opened, a geyser of sea water nearly blasted the men off the deck of the surface ship.

During a dive, Beebe and his companion looked out of two portholes made of ground quartz, seeing things that no man had ever seen before. Beebe, especially interested in the change of light as the sphere descended, describes the scene as "a solid, blue-black world, one which seemed born of a single vibration—blue, blue, forever and forever blue."[1] He observed fish and plankton and discovered at least one new species. The men were supplied with good air by a very simple system: a bottle of oxygen that they could regulate by hand and a tray of chemicals that drew off carbon dioxide. When the atmosphere was stuffy, they circulated the air by waving palm-leaf fans.

On his deepest dive, Beebe went down over 3,000 feet. Although the bathysphere usually went straight down and up again, Beebe experimented with having the surface ship tow it across the bottom. As it was towed, it swayed back and forth so much that it bounced dangerously on the bottom. What seems incredible now is that anyone would trust his life so far down in a vessel dependent on a steel cable. If the cable had broken, there would have been no way for the sphere to rise to the surface. The weight of the cable alone, as it got longer and longer, presented a danger.

There was great public interest during this period in all kinds of exploration. Men were going to the North Pole and flying over it. They explored the Antarctic, flew across the ocean, went up into the stratosphere, always trying to set new records. Nations vied with each other for the honor of climbing the highest or diving the deepest. Inspired partly by scientific interest and partly by raw courage, men and women accomplished new feats that the newspapers and magazines of the day were eager to write about. At the time, there were few commercial applications for these new developments.

One man first turned his attention to the stratosphere and then used the same principles to explore beneath the water. Auguste Piccard was a Swiss scientist already known for the invention of a number of instruments for taking accurate measurements. In 1931, using a gas balloon and a round, airtight steel cabin, he and a companion ascended nearly ten miles into the stratosphere. Just before World War II, Piccard designed a vehicle that could descend into the sea. He called this a bathyscaphe. Not until 1948 was it constructed and tested.

The bathyscaphe worked on the same principle as the balloon except that it started by going down and finished by coming up, instead of the other way round. The equivalent of the balloon gondola in the bathyscaphe was a strong watertight steel sphere that carried the crew—usually two men—and had to withstand deep ocean pressure. To withstand this pressure, the sphere was heavy, its weight exceeding its buoyancy. This negative buoyancy was balanced by mounting it beneath the float—a large metal tank, shaped like a stubby submarine. The float was filled with gasoline, which is lighter than water, just as the gas in a stratospheric balloon is lighter than air.

When a small amount of gasoline was released, the bathyscaphe became heavier than the water it displaced and, therefore, descended. In order to return to the surface, the bathyscaphe carried iron-shot ballast which was held in tanks controlled by electromagnetic valves. The weight of the gasoline was carefully calculated so that its positive buoyancy balanced the negative buoyancy of the total vehicle, including the float, the sphere, the two men, the iron ballast, instruments, and equipment. Floodable air tanks provided additional buoyancy so that the bathyscaphe floated high enough when on the surface.

The men in the sphere could start the descent by opening diving valves to flood the air tanks and select the desired rate of descent by venting the necessary amount of gasoline. The rate of descent was reduced by releasing shot. Return to the surface was also accomplished by releasing shot. In case of electrical failure, the electromagnetic valves would automatically drop the shot and the bathyscaphe would return to the surface. In extreme emergencies, additional weights—such as the electric batteries—could be dropped to assure return to the surface.

Small motors were used to maneuver on the bottom, but

the bathyscaphe did not have power enough to travel any great distance. It was either carried to the place of exploration on a mother ship and then filled with gasoline, or it was filled first and towed to the site. One difficulty, especially with the first model Piccard built, was its inability to withstand heavy weather on the surface.

Piccard's stratospheric balloon had been called the *FNRS* after the initials of the Belgian scientific organization that financed it (Fonds National Belge de la Recherche Scientifique). The first bathyscaphe built by Piccard was backed by the same organization and called the *FNRS II*. It was tested off Dakar in 1948 and made its deepest dive with an automatic pilot to 770 fathoms (4,620 feet). Piccard intended to follow the remote-controlled dive by a second test with himself as pilot, but bad weather forced him to cancel further experiments. The *FNRS II* laid the groundwork for the bathyscaphe *Trieste,* built by Piccard and his son Jacques in the 1950's, which set a record for depth of dive.

Unlike Beebe, Piccard wanted to build a machine that could be improved and developed until it became an acceptable and normal means of descending into the sea—a deep-sea vehicle with nothing in its construction to prevent it from going down to any possible depth. Like Beebe's sphere, Piccard's bathyscaphe moved principally up and down, but it could also hover—that is, remain for periods of time at any depth between the surface and the bottom. It had some ability to turn freely in any direction. The introduction of portholes in both vehicles was a major step forward. Both ships had been built with private capital, although Piccard tried to work with the French Navy at one point and eventually sold the *Trieste* to the United States Navy. There was no commercial use for either vehicle at the time. Interest was chiefly in their ability to aid scientific research.

Auguste Piccard believed firmly in the future of submarine vehicles for everyday use. He thought that passengers and freight could be carried across the ocean in vessels traveling just under the surface. Such ships would escape air drag, have reduced hull resistance, and be completely immune to rough weather. Piccard felt that ships might be designed like dolphins, which react to every force in the water, following a smooth path instead of churning up a wake.

When Beebe and Piccard descended into the water, no one raised any question about the ownership of the ocean in which they worked. No one cared whether or not they were in territorial waters, and there was no suspicion that they might be prospecting for valuable minerals. As scientists on scientific missions, they were free to dive where they pleased. Only later did the question arise whether scientific work cloaked hidden objectives.

While undersea vehicles were being developed in the early twentieth century, diving moved in a new direction. This came about with the use of scuba gear. Scuba means Self-Contained Underwater Breathing Apparatus, in contrast to the helmet-and-hose diving equipment that kept a man tied by his air supply to a ship on the surface. With self-contained apparatus, a diver was able to move freely under water, limited only by the depth to which he could go and the length of time that his air supply lasted.

By 1900, many basic principles of scuba were known, although equipment was quite primitive by present standards. A little later, navies experimented with similar diving gear that might enable men to escape from damaged submarines.

Except in such special cases there was little incentive to develop self-contained diving equipment, for helmeted divers were being used successfully for commercial and military purposes.

By 1926, a Frenchman named Yves Le Prieur became interested in perfecting self-contained diving gear. He knew that helmeted divers could stay down for long periods of time, but had no freedom of lateral movement. On the other hand, divers without equipment could move freely back and forth, but could stay submerged only a short time. Le Prieur tried to combine the advantages of both systems. He and an associate patented an apparatus that included a steel cylinder of compressed air, a short air hose, and a mouthpiece. The whole thing weighed about twenty pounds and enabled a diver to stay down about ten minutes.

Later, Le Prieur improved his apparatus. A face mask took the place of goggles, and the air supply allowed much longer dives. At about the same time, swim fins were invented. Exactly who first devised them is not known, but credit is given to another Frenchman, Louis de Corlieu. With fins and scuba gear, a diver could move freely in the water in a normal swimming position. He was not weighted at the feet and was not limited to moving up and down.

Diving, called "goggle diving" at the time, became very popular in the 1930's. Especially in France, men such as Jacques-Yves Cousteau became adept at the sport. When men were needed for new underwater military purposes in World War II, governments did not have to begin with inexperienced recruits. A small nucleus of men trained in diving already existed.

There are two basic kinds of scuba gear: open-circuit compressed-air apparatus and closed-circuit oxygen gear. In a closed-circuit system, air breathed by the diver is channeled back through some kind of purifying material so that the remaining oxygen can be used again. In an open-circuit system, the used air is exhaled into the water, leaving a trail of bubbles.

Le Prieur's breathing apparatus had one major defect: air was allowed to reach the diver's mouth in a steady, uncontrolled stream. Unused air escaped into the water, a wasteful process considering the small air supply a diver could carry on his back. During the war, Jacques-Yves Cousteau attacked this problem and, with the help of Emile Gagnan, found a workable solution.

Gagnan, who worked for a commercial gas company, adapted a plastic valve already in use to produce a demand intake valve for a diver's mouthpiece. This allowed a diver to regulate the amount of air coming in, according to his needs. The whole system was an open-circuit one, which meant that bubbles escaped into the water. It was not well adapted to military purposes, where concealment is considered desirable, but was eminently practical for peacetime use. In military use, the closed-circuit type is preferred because it allows divers to escape detection from surface observers.

Although both air and oxygen limit the depth to which a diver can go, compressed air is often chosen for peacetime use because it is readily available—as close as the nearest pump—whereas oxygen must be purchased. The combination of helium and oxygen that now allows divers to go to greater depths was used as far back as 1935 but did not come into widespread use until much later.

In 1943, the new Cousteau–Gagnan apparatus was used in

the Mediterranean by Cousteau's friend Frédéric Dumas in a dive that reached 220 feet and lasted fifteen minutes. As soon as the war was over, in 1946, the new device—now called the Aqua-Lung—appeared on the civilian market. It made diving fairly cheap and available to thousands of people. Although Cousteau was not the actual inventor of scuba apparatus, his was the first equipment that was practical, cheap, and easy to use. Variations have been introduced in diving gear since the first Aqua-Lung, but few new principles have been discovered.

Jacques-Yves Cousteau seems to have been present wherever new underwater events were taking place. He helped in at least one archaeological exploration and was present when Piccard made his first dive off Africa. Later, he became one of the first men to attempt the building of a habitat under water and to create a submersible used in connection with it. Perhaps one of Cousteau's greatest contributions has been to make the general public aware of the underwater world. He has used underwater color photography to create full-length motion pictures that have impressed a wide audience and has helped to produce an atmosphere of interest and encouragement for all undersea efforts.

During World War II, diving made great strides within the military establishments. The Italians took the lead in developing a new kind of underwater warfare—the use of scuba divers riding torpedoes or other types of chariots to attack enemy ships beneath the surface of the water. Traditionally, underwater warfare is said to be the weapon of the weaker side. Perhaps no longer true today, it was true of the Italians in World War II, who had no large surface fleet to combat the superior strength of the British in the Mediterranean. Yet, between 1941 and 1943, the Italians were credited with sinking sixteen British ships, two of which were battleships. Other nations developed teams of divers to attack ships and to perform defensive tasks such as underwater demolition. The typical "frogman" with his rubber suit, swim fins, and bizarre face mask captured the public imagination during and after the war. Technical advances were made in the manufacture of all elements of scuba gear, and the war produced trained and experienced divers who returned to civilian life and spread interest in diving.

In the years after the war, enthusiasm for diving grew, and diving clubs sprang up in many parts of the world. While helmet-and-hose divers were seldom used except for one or two types of work, it became routine for many industries to use skin divers in their operations. Advances in commercial diving and sports diving went hand in hand. Equipment became cheaper and facilities for servicing it were found almost everywhere. A new world had opened up.

There was little question at first about how or where divers should dive, but there was concern from the beginning about protecting divers from themselves. Since scuba gear is not entirely safe for anyone not properly trained, diving clubs became interested in establishing safety standards. Concern also developed about divers who speared fish wantonly with no interest in conservation or the need of fish for food. At first, divers gloried in the ability to spear everything in sight. Diving clubs have been instrumental in establishing the idea that spearing fish is both harmful and unsporting. The use of a camera to hunt fish has replaced spearing in popularity. In many places, laws have been enacted that prevent spear fishermen from taking shellfish, such as lobsters. In France, all diving now is licensed, and divers must belong to a recognized club that prevents its members from taking fish and ensures that they use reasonable precautions for safety. Fortunately, divers are seldom able to take fish in commercial quantities, and it has been fairly easy to keep them from competing with commercial fishermen.

Many divers—both amateur and professional—have done underwater archaeology. Divers with no equipment at all brought up ancient objects long before scuba became popular, but new equipment has increased the amount of scientific work that can be done on ancient wrecks and harbors. In addition to bringing up art objects, archaeologists now try to reconstruct the design of Roman ships and discover facts about cargoes and crews. Many objects under water have been covered by sand that has preserved them throughout the centuries. Scientists consider it important not to disturb historic sites until they can be investigated carefully. The Mediterranean is one of the most popular locations for archaeological diving.

Archaeology is controlled by special rules that developed on land and have been extended to the water. Many sites of ancient wrecks are close to shore because ancient mariners tried to remain near land. For the most part, divers have to get permission from the coastal country to begin exploration in its waters. Many nations with historic pasts have built up, through the years, special laws regarding art objects found on their land. These rules have carried over naturally to things discovered offshore. Greece, for example, has very strict rules about removing any art object from the country, and underwater archaeologists must work within those rules. There has been question about amateurs, who— untrained in the cautious techniques of the professional archaeologist—disturb sites in a search for souvenirs. Many diving clubs are trying to make their members aware of this threat to historical research. Few international questions have been raised about archaeological diving, for it usually takes place within territorial waters, often inside harbors. Permission will probably always be granted for such work on terms similar to those that control archaeological work on land.

Along with major advances under the ocean, a number of improvements in secondary techniques have taken place. Many inventions first used on land were soon tried out in the sea. Experiments with waterproof cameras took place in the 1890's, and such cameras were constantly improved during the next fifty years. Before World War I, movie-maker John Williamson shot film underwater by going down in a cabin hung beneath a barge. His precedent-setting movie of Jules Verne's *Twenty Thousand Leagues Under the Sea* was a great success in 1915. Techniques in both still and moving pictures became more sophisticated in the following decades. Cousteau, with his new diving gear, took cameras into the water and, with great patience and skill, made some of the best early scientific movies. After World War II, underwater color photography became very popular. If cameras were to be used far below the surface, lights had to be perfected to accompany them. The strobe light, used to photograph very fast, minute action, such as the falling of a drop of liquid, was found to be excellent under water.

A number of other useful tools became available: waterproof watches, throat microphones, slates for writing. But the most important development during this period was probably the use of underwater sound devices.

The word *sonar* was first used to mean a sound instrument for detection of submarines, but has come to stand for any kind of underwater sound device. The principles of the device, and the first experiments with it, were worked out before World War I by a French physicist, Paul Langevin. During that war a kind of sonar, the hydrophone, was used to detect enemy submarines.

There are two basic kinds of sonar. The passive type, of which the early hydrophone was one example, merely picks up sounds emitted by other objects, such as a moving submarine. The second type of sonar is active: it sends out a series of sounds that are reflected by objects in the sea, return, and are picked up by a listening device. Since radio is not very useful beneath the water, sonar accomplishes many necessary tasks. Refinements have been made in sound-echoing devices so that they can now do many complex things, and variations on the sonar principle are used in almost every form of underwater activity.

Echo-sounding devices are used in a number of ways. Modern fishermen scan the bottom for irregularities that might indicate the presence of fish. They can hear sounds given off by an individual whale or by a school of small fish. Oceanographers use sonar to make contour maps of the bottom, to estimate its composition, and to listen to the squeaks and grunts that make up the language of fish. Oil men use seismic reflections, which are sound echoes, to help them find oil. Three major methods are used in locating oil deposits: the measurement of magnetic forces, the measurement of variations in gravity, and the seismic-reflection method. To make a seismic survey, scientists cause an explosion under water and listen to the sounds that travel down and are reflected back from the various layers and irregularities in the rock.

Probably the biggest change in our thinking about the

sea has come about with the discovery of petroleum deposits in submerged lands offshore. In terms of money spent, of yield expected, of importance to national defense, and of effect on our everyday lives, offshore oil is probably the biggest new resource discovered in the sea. The fishing industry has always had great influence on the laws governing the use of ocean resources, and food supplies are of primary importance in a world of exploding populations. But in modern industrial societies, petroleum is nearly as important as food.

The development of atomic energy has taken some pressure off competition arising from the need for oil. But before and during World War II, no one envisioned this new fuel source. When petroleum was found under the seabed, there was immediate concern about who owned the ocean floor. It requires enormous capital to prospect for oil. This capital does not usually become available without assurances about the ownership of a future well.

Oil reserves are not important merely because they may supply profit for individuals; petroleum is one of the resources on which the whole nation depends. No country owns so many oil deposits that it can ignore this new supply of mineral wealth under the sea. Fortunately for the development of international law, not every nation has oil supplies off its coast, and few countries have the capital and technology for oil exploitation. As a result, when offshore oil was first discovered, only a small number of nations were involved in claims over its ownership. One advantage in such legal disputes is that oil—unlike fish—stays in one place and does not migrate. There is always the question, of course, of two oil wells, on the two sides of a legal boundary line, both tapping the same underground source of oil. But this is a simple question compared with the complexities of legal agreements that involve migrating fish.

No single time or place marks the first recovery of offshore oil. The process was gradual. Men discovered that pools of oil under beaches and headlands extended under the water. They found that rock formations beneath the land continued, far from the shore, into the continental shelf. They could predict that oil reserves existed there as well as on land. Ways were invented to drill at an angle to reach oil not directly under the driller's rig. Perhaps this technique was developed to help one oil man drill diagonally into supplies of oil under his neighbor's property, and disputes have occurred over this question. But the technique can also be used to drill from a beach out under the water.

The earliest real drilling in water took place in the swampy areas along the Louisiana coast and in the shallow waters of Lake Maracaibo in Venezuela. Oil prospecting kept pushing into the wet lands. Barges were used in marshland, the oil rigs being operated over the side. To the present day, the main body of an offshore oil rig is often called a barge, although it does not look like one. It was found to be no more difficult to implant the feet of a drilling rig or an oil well in shallow water than on dry land. In California the oil derricks seemed to march down the beaches, across the land between high and low tide, and straight into the water.

In the United States, the states, acting on the assumption that they owned the submerged lands off their coasts, began to issue leases to oil prospectors. At the same time there was a movement to encourage the United States Navy to claim the oil off the coast of California as a military reserve. To justify this, the federal government had either to prove its ownership or to pass laws making it the new owner. The situation was complicated by the fact that wells on shore, some of them drilled diagonally, were already tapping the same source of oil that the Navy wished to use as a reserve. (The issue of state versus federal ownership of submerged lands is discussed more fully in Chapter IV.) Nevertheless, the question of oil off California brought the legal implications of the ownership of the ocean floor to the attention of the public.

Interest in offshore California oil and the enormous wartime need for petroleum prompted President Truman to state United States policy on submerged lands. In 1945 he issued a proclamation entitled *Policy of the United States with Respect to the Natural Resources of the Subsoil and Sea Bed of the Continental Shelf*.[2] This statement was concerned with the offshore areas outside the three-mile territorial limit but still on the continental shelf.

Simultaneously, Truman issued a proclamation about fisheries in the same area. It stated that fisheries resources developed by United States citizens and preserved by their conservation efforts should be respected by other nations, even when the fish migrated to the high seas. The two proclamations, issued side by side, caused confusion. In the following years, various nations interpreted the two statements according to their own needs and used them together as a precedent for their claims. The effect of the proclamation on natural resources was increased by the fact that nations with fisheries, but without oil, made no distinction between the seabed and the waters above it.

The proclamation on natural resources lists the facts that created a need for this statement of policy: oil and other minerals exist in offshore submerged lands, and it will be possible to develop them in the near future; there is a long-range need for new supplies of these resources; it is important to use and conserve these supplies; for this purpose, some control is necessary, and it is reasonable for the nearest coastal nation to exercise this control; the continental shelf is an extension of the land along the coast; for purposes of self-protection, a coastal nation must keep close watch over the activities off its shores.

Therefore, the proclamation says, the United States regards the natural resources of the subsoil and seabed of the continental shelf beneath the high seas, but contiguous to the United States, as under its jurisdiction and control. The seabed in question is clearly along the coast, but outside the three-mile limit. Truman used the words "jurisdiction and control" rather than the stronger word "sovereignty," a point that lawyers have been arguing over ever since.

At the time of the proclamation, the White House issued a press release explaining that the new policy did not concern arguments about federal versus state control over oil lands such as those off California. However, some people suspected that Truman's proclamation putting oil outside the territorial limit under federal control subtly implied that oil inside the territorial limit should also be under federal control.

Truman's 1945 proclamation on natural resources had a decided effect on world opinion about ownership of the sea, but confusion arose over the difference between natural resources on the ocean floor and fishing rights in the waters above. Many nations, using Truman's proclamation on natural resources as precedent, acted on the assumption that, if Truman could claim control of the seabed to the edge of the continental shelf, they could claim fishing rights and extend their territorial waters farther than the traditional three-mile limit.

Other nations with mineral resources along their coasts issued claims to the seabed of the continental shelf. Nations on the Persian Gulf took control of the seabed along their coasts and divided the entire area by setting up boundary lines agreed to by all the nations concerned. Some British colonies, and Latin American countries such as Honduras and Brazil, claimed the soil of their continental shelves. In 1953, Australia took jurisdiction not only over mineral resources but also over sedentary fish attached to the seabed. Because this included the pearl and oyster industries, the action led to a dispute with Japan.

The greatest problem has come from countries that either extended their territorial waters or claimed fishing rights in the whole body of water over the shelf. Some did not interfere with navigation, but others—for example, Chile, Ecuador, and Peru—claimed complete sovereignty over a zone stretching two hundred miles out from their coasts.

Thirteen years after President Truman's proclamations, coastal ownership of the seabed of the continental shelf was formally established in international law. The 1958 Geneva Conference produced the Convention on the Continental Shelf, which declares that mineral resources of the shelf belong to the adjacent coastal nations. Living resources, such as lobster, are included in control of the shelf, provided that, when harvested, they can move only in contact with the bottom. All the rights of the high seas, including navigation and access to free-swimming fish, remain open to all. However, nothing was established at that time about ownership of the deep ocean floor beyond the continental shelf.

Ocean Resources: Oils and Other Minerals

13 Aerial view of Maracaibo, Venezuela, one of the first places where oil drilling in water took place

14 Jack-up rig digging for oil in the Gulf of Mexico

15 Oil from a "blowout" surges to the surface around a drilling rig off the California coast near Santa Barbara

16   Workmen in small boats, trying to keep ahead of the oil slick, scoop up oil-soaked hay in Santa Barbara harbor

17   Workmen also use pitchforks, rakes, and shovels to pick up oil-soaked hay

18   A log boom is towed toward the entrance of Santa Barbara harbor in an attempt to stop the oil flow

19 Off-shore oil rig "Sea Gem" before it capsized

20 British drilling rig "Sea Quest" which replaced "Sea Gem"

21 A potash plant in Israel extracts minerals from the Dead Sea

22 A gold miner surfaces as his colleague tends the sluice box. The gasoline engine on the raft at left supplies the power for the vacuum dredge that sucks the material from the river bed

23 Prying the baffles from the bed of the sluice box. Everything brought up by the vacuum dredge is washed across the baffles, which trap the heavier dirt and rocks, among which any gold would be found

24 The contents of the sluice box are strained into a large washtub to remove the larger rocks, which are checked, before being thrown out, to make sure none are gold nuggets

25 The final step: old-fashioned gold panning to check what's left in the washtub

CHAPTER 4

Oil and Other Minerals

One day in 1937, Harold L. Ickes, Secretary of the Interior, was having lunch with President Franklin D. Roosevelt, one of the President's advisers, and the President's son James. At first the four men talked generally about travel and fishing. Then, quite casually, Roosevelt asked who owned the underwater lands inside the three-mile limit off the coast of the United States. Ickes answered that he did not know but would ask his lawyers.

This was the first time the question had been raised at a high level of government. Did the lands belong to the adjacent states or to the federal government? The argument over this question lasted more than twenty years. Congress held numerous hearings on the subject. The front pages of the country's newspapers carried the story. Tempers ran high, and important government officials resigned their offices over the issue. It involved the principle of states' rights against the power of the central government, and concerned claims worth millions of dollars.

At first the dispute was called the tidelands oil controversy, but lawyers later used the term "submerged lands," which more accurately describes the area. Technically, tidelands extend only from the high- to the low-water line, but the seabed involved in the argument stretches from low water to the territorial limit. Later, the seabed far beyond territorial waters became valuable because it also was found to contain oil.

Before drilling an offshore well, an oil company must lease the land from the owner, who may charge the company a rent that is reasonable in comparison with the value of the oil that may be discovered; in some cases the owner also collects a royalty—a percentage of the value of the oil produced. From submerged lands off the United States coast the owner—either a state or the federal government—would make a lot of money.

Until the discovery of oil in offshore lands, the states assumed that they owned the seabed along their coasts. They regularly issued permits for people to build docks or to reclaim land beyond the low-water line. Louisiana, in 1933, passed a law stipulating that territorial waters off her coast were twenty-seven miles wide instead of the usual three. Because no one questioned the states' ownership of their territorial waters, the states felt free to issue permits and pass laws concerning them. The United States federal government also acted as if the states owned the land. The Army Engineers had to decide whether new structures in the water hampered navigation that was necessary to national defense, but ownership of the seabed was not involved. The Department of the Interior appeared to accept state control of submerged lands, for, when asked to issue a lease or permit for these waters, the Department asserted that the federal government was not concerned in the matter.

The prospect of great oil resources under the sea changed the offshore oil controversy from a small legal question to a dispute of national importance. The states wished to continue their control of submerged lands. Money from oil leases could be used for schools and roads inside the state. Many Congressional leaders, including some from inland areas, supported state control because they believed that the federal government should not interfere with states' rights. Any effort by the government in Washington to take over submerged lands was looked upon as a land grab, a power play by big government, a step toward socialism.

In contrast, many lawmakers believed that oil resources, if under federal control, could be conserved for the future rather than used for short-range gains. They felt that collection of revenue by Washington would benefit all states equally. Certain oil men sided with this view because they believed that oil leases would be cheaper in Washington than in the individual states.

At first Congress dealt with the oil off the coast of California as a single issue, distinct from the general problem of the other coastal states. But most people realized that a law affecting California would be extended to other parts of the United States. One of the greatest difficulties was that no one seemed to know what the existing law was. Opinions differed widely.

Roosevelt's Secretary of the Interior, Harold Ickes, was a leading figure in the tidelands oil controversy. A man of great integrity, he was famous for his sharp tongue and his persistence in a fight. He held strong opinions and believed passionately in the conservation of natural resources. Private companies, Ickes felt, were greedy about using up national wealth, and he sometimes said that all mineral deposits in the country should be put under federal control in order to conserve them. Because Ickes believed that only the federal government could keep the offshore oil resources from being used up foolishly, he became a moving force in having the federal government claim the submerged lands.

In 1938, Ickes helped a North Dakota senator introduce in Congress a bill giving the submerged lands to the federal government, but it was withdrawn when Congressional lawyers declared that the federal government already owned the land: Congress could not give the government something it already owned. This idea was then introduced as a resolution that urged the federal government to take steps to make its claim good.

Many conflicting opinions gave rise to new forms of the resolution and to Congressional hearings. When the Department of the Navy suggested that the oil off California be held as a military reserve for use in case of war in the Pacific, the question arose whether national submerged lands would be under the Navy's jurisdiction or under that of the Department of the Interior. Nothing had been settled when World War II broke out and directed attention elsewhere.

During the Truman administration, Ickes continued to urge the federal government to assert its claim to submerged lands by suing the state of California, but Congress opposed this. Legislation that favored state claims was introduced, contrary to the resolutions discussed before the war. Congress even passed bills in which the federal government was to give up all claim to offshore lands—total victory for the states—but Truman vetoed the bills. Additional Congressional hearings resulted in widespread publicity. Ickes resigned, partly because of bitterness over the battle.

In 1946 the federal government took California to court for issuing oil leases to offshore lands. The case of *United States versus California* was followed by lawsuits against the other coastal states having offshore oil. In each case the court declared that the federal government owned the submerged lands and that the states were to give up any right to collect money from oil leases.

During this period of controversy, the oil companies had hesitated to drill in submerged lands because of their uncertainty about the law. If they paid money for oil leases, they wanted to be sure that their claims would be honored. After the court's decision in favor of the United States against California, exploration for oil began in earnest, with the federal government in control.

Then Eisenhower was elected President. Because of what had happened in Texas, he believed that lands near the shore belong to the states. When Texas was an independent country, it had claimed territorial waters of slightly more than ten and one-third miles. When it became a state of the Union, Texas offered this offshore land, along with other public holdings, to the government in Washington, provided the government would pay the state's public debt. Congress rejected the offer. Eisenhower argued that, because the United States never paid Texas' debt, the state must still own the ten and one-third miles of offshore land.

With Eisenhower's endorsement, Congress passed two bills

that have almost settled the question. The Submerged Lands Act of 1953 gives the individual states ownership of offshore lands out to the territorial limit. The Outer Continental Shelf Lands Act gives all land beyond the limit, but still on the continental shelf, to the federal government.

This has turned out to be a just and reasonable division. It was thought at first that most of the oil would be taken from areas close to shore; but, with the development of new techniques, oil began to be recovered from the deeper parts of the continental shelf.

The controversy, however, continues. Since the Submerged Lands Act of 1953 did not say precisely how wide territorial waters were for each state, the question returned to the courts. Texas was given the ten and one-third miles she had claimed as an independent nation. Florida, on its west coast, has the same width. California, Louisiana, and Alabama were each awarded three miles, but Louisiana continues to claim a much wider territory.

Legal arguments also continued about the line from which territorial waters are measured. This base line, marking the end of the land and the beginning of the sea, is placed at the low-water mark along an even beach, but must be drawn at some point across the entrance to a harbor or bay. When oil leases are to marked off on a chart, it is important whether the base line falls at the inward or the outward end of a long jetty, whether or not it includes rocks that are awash, whether it crosses a harbor ten miles or twenty-four miles in width. Legal opinions about the boundary of the land have become infinitely complex. Some oil men believe they will never live to see the end of court disputes about oil leases, but perhaps their grandchildren will.

Legal disputes, however, no longer hold up the process of

drilling. When there is doubt about ownership the federal government collects money for leases and holds it in escrow (for safekeeping) until the courts decide to whom it goes. The scale of offshore oil activities and the money invested in them are staggering. In non-Communist countries, offshore wells account for 17 percent of the oil and 6 percent of the natural gas produced. Since 1946, more than ten thousand wells have been drilled in United States offshore waters, and more than 13 billion dollars have been invested in petroleum exploration and development. From this activity the federal government has collected 3.4 billion dollars in oil-lease income. The total amount of capital invested by oil companies is justified, from their point of view, by the amount of money they expect to get back. There seems to be no slowdown in the continued search for new oil reserves.

Oil-well structures standing in the water create numerous legal difficulties, for there is no worldwide agreement about the status of a floating platform twenty miles offshore. Is it an island in national territory or a ship at sea? The British maintain that it is a ship and require the crewmen to carry passports in order to return to port. But, if it were a ship, the senior officer would have the rights and duties of a ship's captain, and he does not. The United States tends to treat oil-well platforms as islands, especially inside the territorial limits where state law applies.

Oil rigs conflict with regular navigation and other uses of the sea, although their position is carefully controlled by law to prevent interference with established sea lanes or harbor entrances. The law requires them to have foghorns and beacon lights, but occasionally a platform is abandoned in the middle of the water without lights or other warning devices. Even rigs in use sometimes fail to comply with safety regulations, because a single foghorn can cost as much as $15,000. Shipping companies complain that oil-well platforms are a hazard to navigation, to which the oil companies reply that ships should not wander off course. In 1965 a freighter going sixteen knots, with its radar turned off, left the shipping lane and ran into an offshore rig. In that case, the court awarded damages to the petroleum company.

Drilling for offshore oil has sometimes caused trouble because of leaking oil. Frequent minor oil spills, which newspapers seldom report, disturb fishermen who fear their effect on nearby fisheries. Extensive damage done to beaches and wildlife by a leaking offshore oil well in 1969 outraged residents of Santa Barbara, California. This was followed by a spectacular fire on a platform off Louisiana, accompanied by a major oil spill. Efforts to clean up polluted beaches are costly and not always effective. Petroleum breaks down in the sea by bacterial action and eventually disappears, but some of the chemical agents used to treat oil spills are long-lasting and create greater problems than those they are supposed to solve.

New equipment has been designed for offshore oil exploration. The drilling is more complicated than drilling on land and may cost ten times as much. Platforms often cost as much as 10 million dollars to build, and $10,000 a day to operate. No two of the larger drilling rigs are exactly alike. Used to drill the original hole that finds oil, these elaborate structures, because of their great cost, are moved as soon as a well has proven successful. A proven well may be capped and held in reserve, or the oil may be led to permanent production platforms and from there to pipelines or tankers.

Some rigs used in shallow water have feet that rest on the

bottom. During towing, the legs are raised in the air; once the rig is in position, they are jacked down into the water. The platform, at the same time, is raised on the legs until it stands high above the sea. The length of the legs can be adjusted to meet any irregularity in the ocean floor.

A newer type of drilling rig is partially submerged but does not rest on the bottom. Several anchors, put out at different angles, hold the structure in one position with respect to the seabed. By a newer method called "dynamic positioning," motors automatically winch the anchor cables according to the movement of the current and wind, keeping the rig centered over the drill hole. The submerged part of the structure may be connected to the elevated part by thin legs to avoid presenting a broad face to the waves. The buoyant part of the rig, far beneath the surface, is not affected by wind or current, and the platform is above the expected height of storm waves. These semisubmersible structures, owned by companies that rent them to the various oil concerns, are sometimes towed halfway round the world, though they are difficult to tow.

Most oil-well platforms are large, with space for drilling equipment, housing for the crew, and, sometimes, a helicopter landing area. Members of the crew, called roughnecks, are experts at their jobs. They work long hours and spend a number of days on the rig before they are given a rest period on land. They are paid more offshore than on land because of the greater danger, and every effort is made to consider their comfort. Buildings on the platform may contain air-conditioned sleeping quarters, a mess hall, and a lounge for recreation and television. On some rigs the food is provided by caterers who send out both the supplies and cooks to prepare them.

A number of auxiliary vessels and facilities are needed to support a drilling rig. Helicopters transport some of the crew, and two or three boats take turns ferrying supplies and personnel. It is not always easy to transfer men and materials from a rocking boat to a platform standing high in the water. Sometimes a crane lowers a cargo net down to the boat; men scramble into it and are hoisted up like a load of baggage. At night, with lights blazing, an oil rig standing alone in the otherwise dark water becomes an impressive world of its own.

Once discovered, oil is usually brought to shore by underwater pipeline. On a barge, forty-foot lengths of pipe are welded together, wrapped in bitumen and paper, coated with concrete, and lowered over the stern by a continuous process. Laying pipe by surface ship is a complex operation and may be hampered by rough weather in areas such as the North Sea and the coast of Alaska. Although the oil industry would like an underwater vessel that could lay pipe without concern for the weather, such an advance lies far in the future. The industry would also like a vehicle that could inspect pipeline, a job now being done by expensive divers who can go only so deep. There are already remote-control devices that can perform certain jobs on underwater wellheads, and oil companies continue to automate as much of their equipment as possible. In the future, the new submersibles and divers living on the bottom will be of great use in the offshore oil business.

Although the Gulf of Mexico was one of the first and is still one of the most important sources, offshore oil is found in submerged lands in many other parts of the world. Coastal areas as far apart as Venezuela and Nigeria have offshore

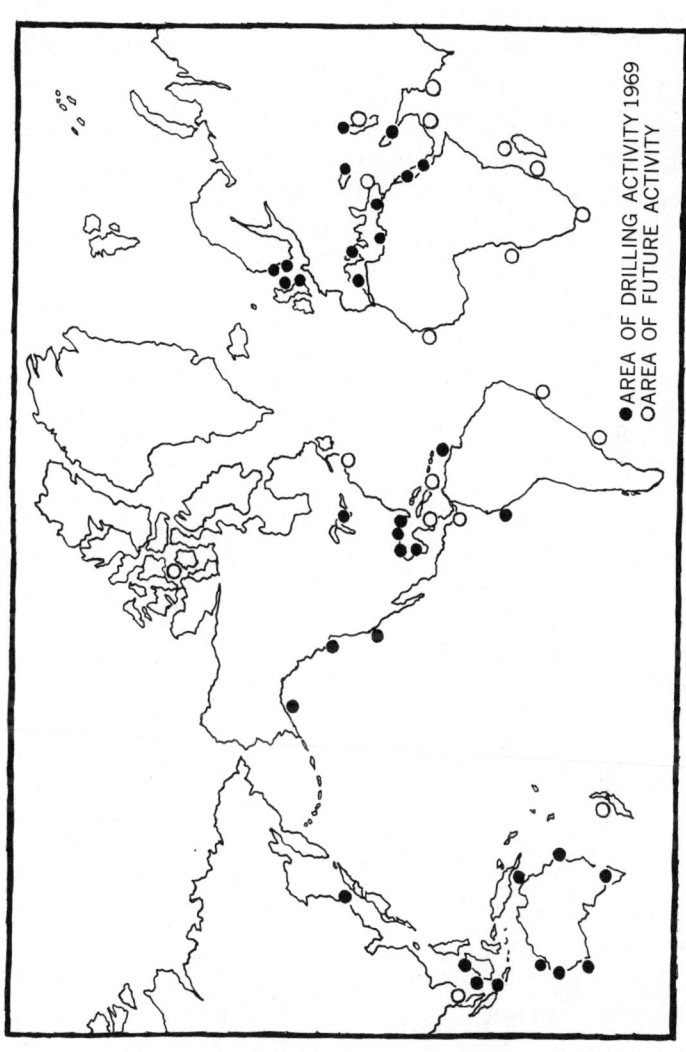

OFFSHORE PETROLEUM SITES—1969

wells, and petroleum has been found near Australia, until recent years a land thought to be completely without oil potential. Oil wells are found off the south coast of Alaska, and the new oil strike on the Alaskan North Slope reaches under the sea. One of the largest supplies of oil is found in the offshore waters of the Near East.

The bottom of the shallow Persian Gulf cannot accurately be called part of the continental shelf, but it is an extension of the land around it. The same nations of the Near East that have profited greatly from oil under their land now also possess offshore oil. In 1957 the first crude oil was pumped from the Salfaniya area, one of the world's largest potential underwater fields, which lies in the Persian Gulf off northern Saudi Arabia. A pipeline connects it to a terminal on the Arabian coast 135 miles away, from which oil is shipped out by tanker. A number of small countries along the Persian Gulf sell oil, mainly to Britain. Bahrein, Abu Dhabi, and Dubia were little more than trading centers in the desert before oil was discovered; Dubia, in fact, was known throughout the region as a great clearing house for smuggled articles such as gold dust, watches, silver bars, and drugs. The oil potential of these countries is small compared with that of such great oil nations as Saudi Arabia, Kuwait, Iraq, and Iran, but it is still useful. Despite controversy among the nations claiming the Persian Gulf seabed, the area has gradually been divided by lines drawn halfway between one coast and the next.

All the political difficulties that beset the oil industry on land also affect the offshore fields. At least two major crises have disrupted Near and Middle Eastern oil production— one in Iran in 1951 and one concerning the Suez Canal in

1956. Both were particularly hard on northern European countries which, at that time, depended heavily on Near and Middle Eastern oil. Oil production is affected not only by political disputes, but by the world price of petroleum. An oversupply of oil in the world since 1957 has driven prices down, although enthusiasm for finding new reserves still continues. Petroleum, however, is not always found in the place where it is needed. Transportation by pipeline and tanker increases its cost in areas such as northern Europe, especially when ships cannot use the Suez Canal. Oil companies are constantly negotiating with the Near and Middle Eastern nations to counteract price changes and political upheaval. For a number of years the companies divided their profits with host nations on a fifty-fifty basis, but a more recent trend is toward agreements that give the countries 60 percent.

One of the most important events in offshore exploration was the discovery of natural gas in the North Sea. Oil and natural gas are formed of the same chemical elements and are often found together. The word "petroleum" is a general term that includes both gas and oil. In early Texas oil fields, natural gas was considered a nuisance and was piped to the top of the rig, where it was set afire and burned harmlessly away. The tiny fires often seen dotting oilfields were natural gas being burned off. Then oil companies discovered that it could be moved about in pipelines and, later, that it could be liquefied and shipped by tanker. Used chiefly in home heating and cooking, it is also a good fuel for industry. Automobiles now designed to run on gasoline could be designed to run equally well on natural gas.

The highly developed areas of northern Europe need large

amounts of fuel, but the coal deposits of England and the continent have been worked for a long time and are no longer sufficient to meet the growing needs. It was believed that there was no hope of finding oil or natural gas in this area until, in 1959, a huge reservoir of gas was discovered in Holland.

This gas field at Groningen, a short distance inland from the Dutch coast, has proved more productive than anyone had thought possible. It promises to fill Holland's need for cooking and heating fuel for some time to come. Its discovery raised tremendous hope that, if this one field existed in northwestern Europe, there might be others like it.

The tense political situation, interference with the use of the Suez Canal, the high cost of transporting oil over long distances, and the uncertainty of oil supplies in time of war—all these made Britain anxious to find a fuel supply nearer home. Natural gas, compressed to a liquid state, was being shipped from Algeria to England, and some oil resources were being developed in Libya and off the west coast of Africa. Although such supplies were less endangered than those from beyond Suez, they nevertheless left England at the mercy of foreign nations. (France was in less difficulty because Algerian oil was being sent by underwater pipeline across the Mediterranean.) Northern European countries, especially England, were ready to spend large sums on the chance that the same type of gas found in the Groningen field could also be found under the North Sea.

As soon as the possibility of offshore fuel was recognized, nations bordering the North Sea began dividing the area into national jurisdictions. In some cases a median line halfway between two coasts was used. When Germany, Denmark, and Holland could not agree on a division, they submitted the dispute to the International Court of Justice. The question has now been returned to the three nations for solution by negotiation.

With their long coast lines, England and Norway together claimed a large share of the North Sea, but at first they had difficulty in agreeing on a division. The floor of the shallow North Sea is part of the continental shelf, which the 1958 Geneva Conference on the Law of the Sea had defined as the seabed adjacent to the shore, to a depth of two hundred meters, or to whatever depth it could be exploited. Along the coast of Norway, close to the coast, the ocean floor is cut by a trench that is much deeper than two hundred meters, and deeper than anyone could drill at the time. If one argued that Norway's shelf stopped at the inner edge of the trench, that country would have almost no offshore land, and England would own most of the North Sea. The question, however, was settled fairly and amicably between the two countries. The deep trench was ignored as a mere irregularity in Norway's continental shelf, and the dividing boundary was drawn halfway between the two coastlines.

As soon as these legal questions had been settled, exploration for oil and natural gas began in earnest. Because of the gas strike in Holland, many oil companies believed that the best chance of success lay in the eastern part of the North Sea, close to the shores of Germany and Denmark. But gas was not found there at first.

Drilling began off the English coast in 1963. There was less hope in this area, but the desire for a local supply of fuel was so great that oil companies were willing to gamble tremendous sums on the possibility of future success. Finally, in 1965, the British Petroleum Company struck natural gas. The drill rig *Sea Gem* made the first British strike. This was a high platform resting on four legs, which stood on the bottom. The site was only forty-two miles from the English coast.

At first the oil company tried to keep news of the gas strike from spreading too quickly. It wanted time to test the well and to make sure that the supply of gas was big enough to be worth exploiting. But it was impossible to keep the news secret for long. The British newspapers were jubilant at the idea that England, at last, had a safe source of fuel. The first well proved to be a good one, with an adequate supply of gas and a location that required only a short pipeline. Indications that other supplies of gas existed nearby have proved to be correct.

The discovery of petroleum under the North Sea may well change the balance of political power in the world. Both the Soviet Union and the United States have large supplies of petroleum, and the Soviet Union has been advancing steadily in the amount of oil she produces. For both nations, offshore oil, being merely an addition to the resources already found on land, does not change their position in the world. In many smaller countries, however, without land resources of oil, any new supplies of fuel are likely to be found offshore. Such fuel increases their political importance, their independence, and their power. Britain and Holland are in a stronger position than they were when most of their fuel had to be imported.

Unfortunately, the gas strike on the *Sea Gem* was followed by disaster. A drill rig remains over a well only until the well has been proven; then it moves on. The *Sea Gem* had finished its job on the first, successful site, and its dismantled equipment was stowed safely away ready for towing. Because it was Christmas, many of the crew had been given a holiday. With no warning, one leg of the *Sea Gem* gave way. Witnesses on a nearby ship reported what happened. There had been no storm, no abnormal wind or waves. One leg of the oil rig buckled, and one corner of the platform tilted, leaning closer and closer to the water. The weight and angle of the platform put too great a strain on the remaining legs, which also collapsed. As the horrified spectators watched, the whole structure slid into the sea.

Of the thirty-two men on board, all but thirteen were rescued. Although the cause of the accident has not been definitely determined, experts think that the sea floor under one corner may have shifted in some way, leaving the leg at that corner without proper support. When it gave way, it pulled the platform over, destroying the other three legs.

In the Gulf of Mexico, hurricanes have taken their toll of drilling rigs. Oil men, accustomed to working on land, at first thought they could survive storms at sea with nothing more than raw courage, but they soon discovered that the terrific wind and waves of hurricanes can destroy equipment that appears safe in normal weather. Petroleum companies now insist on accurate information about wind and wave conditions before they design an oil rig, and they now consider weather forecasting of first importance. Even with such precautions, losses occur. In the past few years, certain insurance companies in London, which insure many of the world's drilling rigs, have doubled their prices and have added special charges for rigs working in hazardous waters such as the North Sea or the Pacific near Alaska.

Besides the oil being developed on the continental shelf, it now seems possible that the deep ocean may provide future sources of petroleum. In 1969 the deep-drilling ship *Glomar Challenger* discovered salt domes and oil-bearing deposits in the deep water of the Gulf of Mexico. The United States Naval Oceanographic Office announced that salt domes along with organic-rich sediment were found in the deep ocean of the eastern Atlantic. Although high costs and lack of technical ability prevent anyone from developing such resources at present, their existence influences estimates of the deep ocean's value.

Oil is the most valuable mineral taken from the sea, but it is not the only one. Rivers pour millions of tons of material into the ocean every year, and many substances that exist on land are found in the seabed also. Sea water itself is made up of a number of elements, some of them in quantities great enough to be useful.

In some areas of the world, the shafts of coal and iron mines, sunk on shore, reach out with tunnels under the ocean bottom. The Japanese mine low-grade iron ore directly from the seabed, separating the metal with magnetic devices and pumping the iron-bearing sludge ashore through a pipeline. They also mine coal by building artificial islands in the sea. One mine shaft, sunk through the center of such an island, brings up coal from two thousand feet below the sea floor.

Valuable minerals can be brought up by dredging in the shallow waters near river mouths, where silt has been washed down from the land. Tin is mined by dredging near the rivers of southeast Asia. Diamonds have been successfully mined off the coast of South Africa. A barge with a long vacuum hose sucks up diamond-bearing gravel from the bottom, passes the gravel over a special surface that holds back the gems, and returns the waste material to the sea. In this way, 2,193 carats of diamonds were brought up in one day.

Seashells, by definition, are not minerals, but they are often handled as minerals. Iceland, which has few natural resources, mines seashells close to shore and uses the material in construction cement. From the continental shelves of the United States, 20 million tons of oyster shells are taken annually as a source of lime. Mother-of-pearl, used for buttons and other bright objects, is gathered off Australia. Small quantities of other types of shell, useful in jewelry and decoration, are also found, but more often on shore than under water.

In certain areas, where a layer of salt lies under the surface rock, a crack or upheaval in the earth causes a core of salt to push upward, producing what is called a salt dome. On land or under water, salt domes signal the possible presence of oil, and a small percentage of them contain another valuable mineral, sulphur. Two mines off the coast of Louisiana currently produce two million tons of sulphur a year.

One of the companies producing offshore sulphur has constructed, at a cost of 30 million dollars, the largest manmade steel island ever placed in the water. There sulphur is brought up in large quantities by hot water pumped down into the mineral deposit. As the sulphur is melted, it is forced to the surface and sent ashore through a heated pipeline. Natural gas, available nearby, is used to supply power to heat the water and run the machinery. All the necessary equipment for handling the sulphur is maintained on the structure, which is almost a half mile long and contains living quarters for 120 men.

The mineral phosphorite, used in fertilizers, is found lying

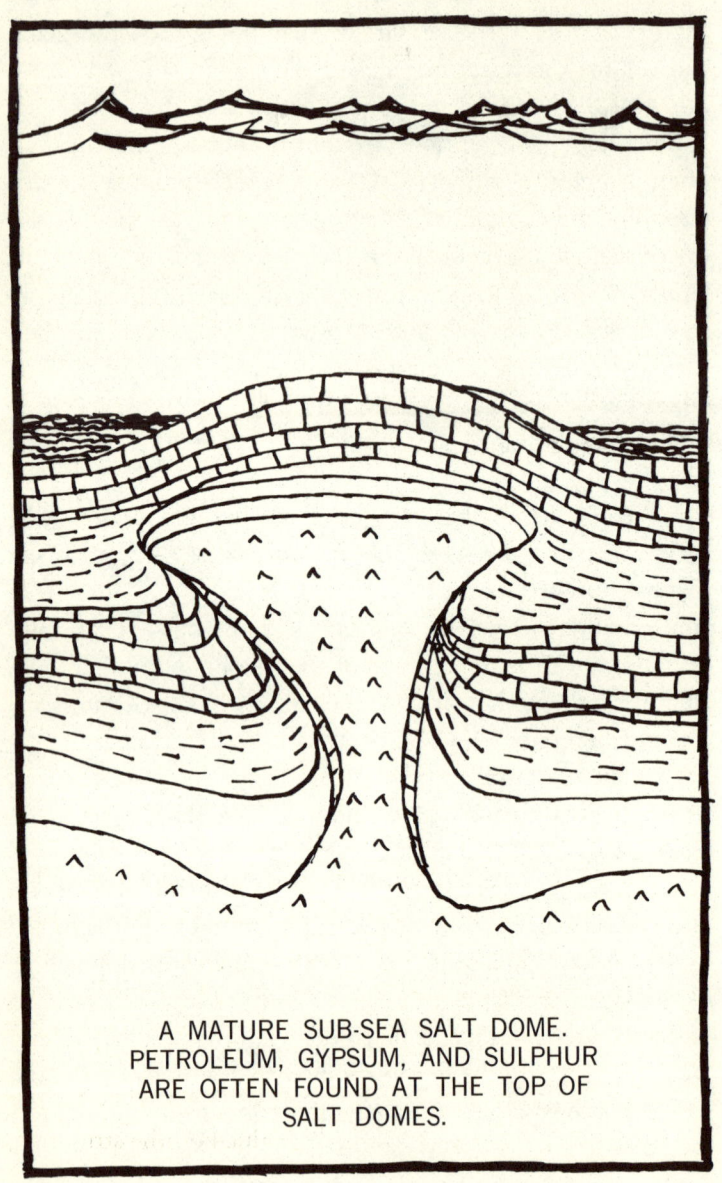

A MATURE SUB-SEA SALT DOME. PETROLEUM, GYPSUM, AND SULPHUR ARE OFTEN FOUND AT THE TOP OF SALT DOMES.

on the sea floor in various-sized chunks. To replenish the soil with phosphorus to make crops grow, farmers of past generations used animal bone, dead fish, and bird droppings on their land. In 1860 scientists discovered that the phosphorite found in rocks can be made into a chemical fertilizer, and then that the phosphorite found on the seabed can be utilized in the same way. There is more than enough phosphorite in the world, but a small number of nations own the largest share of it, and other countries must import it. Relatively cheap to mine, it can be expensive to transport because of its bulk. Accordingly, in certain regions, it may be cheaper to mine it under water than to carry it from one place to another.

One California company, in 1962, paid the government more than 100,000 dollars in rent for a large section of the seabed from which to mine phosphorite. The mineral proved to be there in reasonable quantities, but trial dredging brought up other, unexpected objects. The area had apparently been part of a naval test-firing range, and the sea floor was littered with unexploded mines, live shells, and spent torpedoes. Since such warlike materials could have damaged the dredging equipment or blown up the surface ship, the company reluctantly demanded its money back.

The ocean floor is also strewn with manganese nodules—roundish mineral chunks that may be valuable, for they usually contain manganese, nickel, cobalt, zirconium, and copper. Lying like pebbles on the seabed, some are as small as marbles, others as large as potatoes. Vast areas of the ocean contain these nodules, although their size and composition vary from place to place.

Scientists are not sure how manganese nodules are formed,

but they believe the process continues at a low, steady rate. Many appear to have been built up around a central core of some other material (a shark's tooth or a piece of bone), which remains the heart of the nodule. Scientists believe that inside growth rings—like rings in a tree trunk—indicate the amount by which the nodule increases in size each year.

The question how valuable the nodules are is still an open one. Various pilot projects have studied the feasibility and cost of recovering such seabed minerals as the nodules. Arvid Prado, the United Nations representative from Malta, commenting on the great mineral wealth that lies on the seabed, has expressed the hopes of many small nations for future wealth from the sea. However, Dr. K. O. Emery, Senior Scientist at the Woods Hole Oceanographic Institution, believes that large-scale ocean-floor mining of manganese nodules is unlikely for several decades—at least until the land resources of the metals are depleted and ocean-mining techniques improve. The manganese content of the best nodules is about half of that required for ore to be profitable on land; the concentration of nickel and copper is high enough, but the cost of mining and extracting them would make them too expensive to compete with existing land supplies.

Other mineral deposits have been located under hot brine pools on the floor of the Red Sea—seabed areas that contain appreciable amounts of zinc and copper, and smaller amounts of silver and gold. These hot spots, which remain above the temperature of the surrounding water, might be valuable sources of minerals if the cost of mining were not so high, if extracting useful metals from the surrounding sediment were not so difficult, and if the lands nearby could supply facilities and sources of power.

Mining operations for seabed minerals differ from petroleum exploitation in that they require mobile ships and barges that can move over an area to gather thinly scattered deposits, but they are covered, so far, by the same laws that affect oil installations. Mining in river mouths is directly under coastal control, as are all coastal operations inside the territorial limit. The most available phosphorite and manganese nodules lie on the continental shelf, where they belong to the coastal nation, just as oil and gas deposits do. At some later date, manganese nodules may be gathered from the deep ocean, in which case their ownership will come into question. Although there is much talk about the wealth supposedly available on the deep ocean floor, nations have not yet agreed on rules for ownership of the seabed beyond the continental shelf. In the meantime, it is estimated that the total annual value of all minerals from the ocean floor—except petroleum, sand, and gravel—is not likely to exceed 100 million dollars by 1980.

CHAPTER 5

Ocean Fisheries

Hugo Grotius believed that ocean fish were inexhaustible and, in the first half of the nineteenth century, most men still believed that whales and seals, salmon and herring were like the buffalo of the western plains—so plentiful that they would never diminish. Then fast, power-driven boats came into use, machinery could lift heavy trawl nets, and new methods were invented to preserve the catch. By the 1890's ocean fisheries needed regulation.

As long as fish were plentiful, nations made treaties among themselves, parceling out fishing rights. But under new conditions, three things appeared to be necessary: effective conservation measures, the collection of scientific knowledge on which to base those measures, and a legal structure that would protect fish stocks and divide fishing rights equitably.

International law about fisheries does not rest on any

individual treaty or single document. Each effort at conservation, each decision by a world court, each fishing dispute settled by agreement added some new concept or new method to the management of fisheries.

One of the earliest ways to regulate fisheries was the regional conference which could produce fishing agreements and which often set up a permanent regional commission. In 1881, eight nations met at the Hague to discuss fishing in the northeastern Atlantic, particularly in the North Sea. The conference produced the North Seas Fishing Convention, a formal agreement that aimed at policing the fisheries and preventing the destruction of property. Since that time, a number of conventions and regional commissions have been set up: for the Mediterranean, the northwestern Pacific, the northeastern Atlantic, etc. The commissions act as centers of discussion, collect scientific data, and establish rules for fishing. Their power varies: some are purely advisory, but others produce rules binding on all nations that sign the agreements.

Another method of regulating fisheries is to deal with individual species of fish. Two early conservation efforts dealt with two species of ocean mammals—fur seals and whales.

Far-ranging creatures, fur seals cover great distances in the ocean during their yearly migration to and from the islands where they breed. Every spring they come to the Bering Sea and other parts of the northern Pacific, leave the water, and live for a considerable period on land. The large dominant males collect a harem of females and fight off all opposition from bachelor males that have not won mates. A single pup is born to each female, and the pups remain on the island until they are strong enough to swim.

Fur seals are awkward on land, and hunters in the past often killed them on shore by rounding up a group and knocking them on their heads. With modern guns the hunting of seals became simple slaughter. Seals in the Atlantic were almost exterminated before legal protection saved the remaining herds. Four and a half million seals lived on the islands off Alaska when the United States took over the territory from Russia in 1867. Forty-three years later, in 1910, their number was down to 125,000. More than 97 percent of the animals had been killed off. By the beginning of this century, there was reason to believe that the Pacific fur seals might vanish as a species.

As early as the 1880's, the United States had tried to develop an international agreement to protect the seals, but no positive action was taken, although many nations agreed in theory. By 1911 it was obvious, first, that fur sealing was no longer very profitable, and, second, that if protection were not enforced, no seals would be left. Finally the four countries most concerned—the United States, Canada, Russia, and Japan—agreed to a system of regulation in the Convention for the Preservation and Protection of Fur Seals.

Killing, capturing, and pursuing fur seals on the high seas were outlawed (except for native Aleuts and Eskimos, if they used only primitive hunting methods). Taking fur seals was legal only on breeding grounds such as the Pribilof Islands in the Bering Sea. It was to be done exclusively by the United States and Russia each in its own territory. Canada and Japan each were granted a 15-percent share of the sale of Pribilof Island sealskins in return for not taking seals on the high seas. Now the United States does its sealing through the Fish and Wildlife Service, the only organization allowed to kill seals.

The Fur Seal Convention is unique, for in no other treaty have two nations—in this case Japan and Canada—been paid to refrain from using ocean resources. Writers have speculated about what would happen if other nations, say France and Sweden, also demanded payment for *not* taking seals in the Pacific. Thus far the issue has not arisen. The convention was successful in saving the seals, for by 1941 the seal population on United States islands in the Bering Sea had returned to 1,500,000. The number of seals on Soviet-held islands is still small, but their number was never large.

Obviously the Fur Seal Convention depends on self-restraint by nations that have agreed on a desired result. Because sealing ceased to be profitable when the seal population dwindled, restraint was fairly easy. Because seals come together in one place every year, it has been possible to control them. Although some lessons learned by fur-seal control were used in later treaties, many of the points unfortunately do not apply to fish that swim over great areas of the ocean.

Compared with that of the fur seal, the story of whales is a sad one. Men have been hunting whales for centuries, both for oil and for meat. The great heyday of sperm whaling in the Pacific lasted from 1820 to 1850. Not good for meat, sperm whales yielded excellent oil, used in lamps and candles, and a bone-like material, good for corsets, whips, and umbrellas. After 1846 the industry declined because whales became scarce and petroleum—discovered in 1859—provided a cheap substitute for whale oil.

For a brief period in the nineteenth century, when whaling produced little profit, the remaining whales were relatively safe. Then the use of soap increased, and whale oil became valuable again. By 1900 scientists had discovered how to make margarine that contained some whale oil, and today a palatable spread can be produced entirely from such oil. With the use of refrigeration in ships, whale meat could be brought back to port. The Japanese eat it in quantity but, except in Norway, it has never become popular in other parts of the world.

Since 1900 the whaling industry has been concentrated in Antarctic waters, and ships have hunted a different type of whale from those taken in earlier times. Right whales were a species that swam slowly and floated on the surface when dead. Consequently they could be taken from simple rowboats and were easy to process beside the mother ship. Now modern, wide-ranging vessels with power-driven boats and efficient harpoon guns can take rorqual whales that swim at great speed and sink when killed. Either air must be pumped into the carcasses to keep them afloat during processing, or they must be dragged aboard through an opening in the ship's stern.

The remaining whales in the ocean are endangered by modern whaling fleets that have factory ships, small, fast boats, powerful machinery, and skilled crews. Sonar is used to locate whales and frighten them until they thrash about and become easy to sight. Unlike the fur seals whose breeding grounds lie in territorial waters, whales roam the oceans, and conservation measures to protect them must be made on an international basis.

Efforts to regulate whaling went on for many years, but there was no international agreement to control the industry. There was no central agency to collect scientific knowledge about whales, on which conservation measures must depend. Finally the International Bureau for Whaling Statistics was established in 1930 at Oslo, Norway. It became a focus for the industry and a central clearing house for information.

When prices fell during the depression of the 1930's, England and Norway were able to institute a quota system by which each nation was allowed to take only so many units of whales. The unit is an arbitrary measuring device by which certain small whales count only half as much as one large whale. Then, in 1948, the International Whaling Commission was set up. It limits the number of ships from various nations, establishes the quota of whales that may be taken, and places observers on all whaling vessels. Although the commission has sometimes been criticized as ineffectual, it continues to regulate the modern whaling industry.

Unfortunately, these efforts have not provided successful conservation of whales. The whaling commission has no power to enforce its quotas on nations such as Chile that are not I.W.C. members. Scientific advisers have warned the commission that quotas must be lowered, and certain types like the blue whale protected altogether. But when low enough quotas are suggested, some whaling nations will not accept them. As a result, the blue whales—which numbered 100,000 in 1938—are down to about 600. This is too few to provide much chance for the whales to mate and keep the species alive. The number of fin whales has also been drastically reduced.

Because of the reduction in whale herds, the whaling industry in the Antarctic has ceased to be important for the nations that originally developed it. The British stopped whaling in 1963, the Dutch in 1964. The Norwegians, who sent their first whalers to the Antarctic in 1904, made their last whaling expedition in 1967. The Japanese and the Soviets continue to take whales, but they face ever-diminishing stocks, smaller whales, and lower profits.

Scientists are not optimistic about the future of the remaining whales. Some feel that they will be killed until they become so scarce that it will be unprofitable for anyone to hunt them; perhaps then the few remaining will be able to survive. Other experts feel that since they travel in herds, the whales may become extinct if they are reduced to a number smaller than the herd. There is also the theory that whales may have been headed for extinction anyway because they have a very specialized food supply and have developed to the point of giganticism. In a legal sense, whales appear to be things belonging to no one (as defined in the old Roman law), and no one has taken adequate responsibility for preserving them. They may become as scarce in the Antarctic as they are in other parts of the world.

The United States took an early interest in the conservation of Pacific salmon. These fish, like some other species, are born in a river, swim down to the sea, where they travel great distances, then ultimately return to the river of their birth, and make their way upstream to lay their eggs and die. Until the early part of this century, salmon were more plentiful than they are at present in such places as the Fraser and Columbia rivers, on the northern Pacific coast, and in Alaska. Most salmon were caught at the season when they returned to the rivers, and the most important conservation measures prevented the use of nets at the river mouths during the great salmon runs.

Another danger to salmon is on land where spawning streams must be protected. Blocking of the rivers, as well as pollution, tends to kill off young fish at the source: either the adult fish cannot ascend the river to lay eggs, or the young cannot survive the trip downstream. To overcome such hazards, fishways must be built to bypass dams or other obstructions, and the best possible measures taken against pollution.

Salmon fishing in the northeast Pacific is controlled by a number of interlocking agreements and treaties. The Pacific Salmon Convention between the United States and Canada was ratified in 1937 and has been effective in the management of salmon fisheries. Each nation takes fish in its own territorial waters. The International Pacific Salmon Fisheries Commission makes rules about gear, tries to ensure the maximum yield, and regulates the size of the catch of the two countries. It is also responsible for developing fishways to facilitate salmon migration.

The United States, trying to increase its salmon stock by careful conservation, was concerned about other nations that took salmon in great quantities on the high seas. In 1952, therefore, the United States, Canada, and Japan signed the North Pacific Fisheries Convention, which employs a principle called *abstention*. The three nations agreed that, if a fishery stock is under management and is being fully used by one or more of the countries, other signers of the convention will abstain from fishing that stock within the treaty area. The convention specified, for example, that in certain areas of the Pacific, the United States and Canada would carry out necessary conservation measures and that Japan would abstain from taking halibut, herring, and salmon from those areas. In other areas, the United States was to take conservation measures, and Japan and Canada were to abstain from taking salmon.

According to the abstention principle, if one or two nations put time and effort into conservation, and prevent their own fishermen from depleting a fish stock, other treaty nations should cooperate by not taking the fish that have been saved. Although this principle has not yet been widely accepted, the United States continues to urge that it become part of international law.

Still another building block in the legal structure that surrounds fisheries is the process of submitting disputes to a world court for arbitration. A major dispute between England and Norway was submitted to the United Nations World Court of Justice in 1951 and the decision handed down is a landmark in fisheries law.

For many years English boats had been prevented from fishing in waters near the coast of Norway, especially in the region of the Arctic Circle. While Britain claimed that these were international waters, Norway insisted that they were internal—completely within the boundaries of the country. The difference in interpretation depended on where the base-

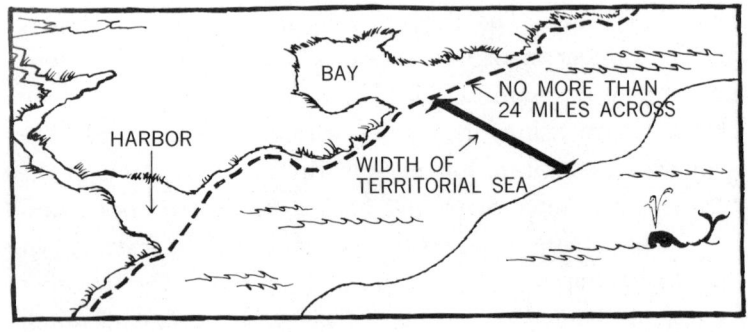

NORMAL BASELINE—ALONG LOW WATER LINE

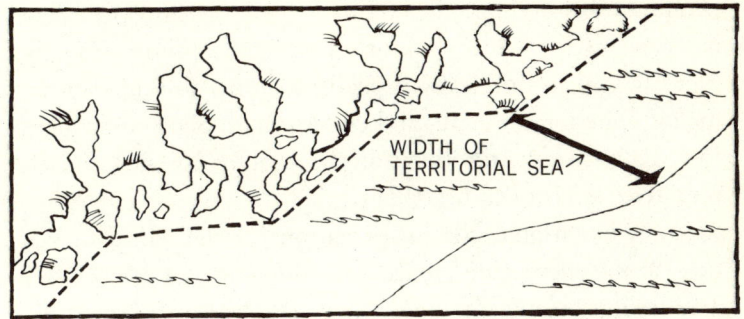

STRAIGHT BASELINE—FROM POINT TO POINT

line was drawn that marked the end of internal waters and the beginning of Norway's territorial sea.

England was in favor of a *normal* baseline that followed the low-water mark along each bend in the coast and was drawn across every bay at a point ten miles wide. Norway claimed that her coast was so irregular, so indented with fjords, and her coastal waters so full of islands, that a normal baseline could not be drawn fairly. Instead she chose a number of points along the coast, many of them on the far side of islands, and drew a *straight baseline* from point to point. Her territorial waters, four miles wide, began at this straight baseline. Such a system of drawing boundaries enclosed a large area of the sea, in which the English were prevented from fishing.

In 1951, the World Court of Justice ruled that Norway's straight baselines were justified for a number of reasons. First, Norway's coast was so irregular that a normal baseline would be difficult to draw. Second, Norway had controlled this enclosed area and used it for fishing during a long period of history. (Such a claim, based on historic usage, has been used by many countries since the ruling.) Finally, Norway was economically dependent on her fishing industry for food, national wealth, and employment. She could, therefore, claim fishing rights inside the straight baseline and for four territorial miles beyond it.

Seven years later, the major nations of the world met at the 1958 Geneva Conference to codify the Law of the Sea. The legal concepts developed over the past years were discussed, and many of them adopted as international law. One of the four documents produced was the Convention on Fishing and Conservation of Living Resources of the High Seas. This establishes as a principle that all fishing nations must take conservation measures to protect living resources, and that such measures must be based on sound scientific knowledge. Conservation measures are defined as those programs which will produce the *optimum sustainable yield*—the greatest amount of food supply that can be taken from a fishery without reducing the stock of fish for the future.

What trends in fishery management have become apparent since the 1950's when the Anglo-Norwegian Fisheries Case and the 1958 Geneva Conference did so much to define the law?

First, the need for scientific knowledge, stressed in the Geneva convention on fishing, has increased, and methods of gathering information have improved. Producing accurate, quantitative estimates of ocean resources is a complex process, which makes it difficult to forecast the future of any fishing stock. But certain types of fish can be counted accurately by echo sounders; the age of fish caught can be judged by the rings on their scales; the distribution of the fish eggs found in water samples indicates the size of fish populations. Statistics about fish catches around the world continue to provide the largest body of information about fishery resources. Since 1945, the Food and Agricultural Organization of the United Nations (FAO) has had a fishery section that gathers data and publishes a yearbook of current information. The need for scientific knowledge is of major importance to fishing law, and efforts to gain greater understanding about fisheries will certainly continue.

A second trend has become apparent since the 1950's: coastal nations are beginning to push their claims farther out into the high seas. Such claims are closely connected with fishing rights, for it is often difficult to distinguish between claims to sovereignty in territorial waters and exclusive fishing zones. Many claims are based on precedents set by the Anglo-Norwegian Fisheries decision. The Soviet Union, partly for security reasons and partly because of competition from Japanese fishermen, closed off all of Peter the Great Bay, near Vladivostok on Russia's Pacific Coast, claiming that this was justified by historic usage. Iceland, claiming economic need, extended her fishing zone to twelve miles. Indonesia, a nation of islands, has set up a straight baseline from island to island, enclosing large areas that would otherwise be international waters.

In 1952, Chile, Ecuador, and Peru (called the CEP countries) made one of the largest claims to offshore waters when they declared that their sovereignty extended two hundred miles off their coast. This includes fishing rights, but is also a direct claim of territorial waters. Since land along the Pacific coast in these three countries drops off steeply and leaves almost no continental shelf, the claim to two hundred miles

of sea may originally have been an effort to secure the equivalent of a continental shelf.

Theories explaining the CEP claims are varied and subtle. The three nations maintain that every country has a right to decide the width of territorial waters, and that their special economic interests justify this width. In the interest of conservation Peru insists that fish off its coast live in a biological environment that extends two hundred miles into the sea. One rather strange reason suggested for CEP claims is that two hundred miles is no greater percentage of the Pacific than the accepted three-mile territorial limit is of the Atlantic. The CEP nations, making little distinction between fishing rights and national security, also maintain that this wide territorial sea is necessary for their military safety. In spite of all these arguments, many nations, including the United States, do not accept the claim to two hundred miles of water.

Peru has good reason to take special interest in fishing rights. In the last twenty years, Peru has changed from a minor fishing nation to one of the world's largest producers of fish. Its chief catch is the anchoveta, a small fish previously considered of value only as food for birds. Now Peru, with a greatly expanded fishing fleet, brings in large quantities of anchoveta to be processed into fish meal. Previously used only as fertilizer, fish meal has been found to be an excellent food for poultry and hogs, if the percentage of meal in the total diet is carefully controlled. Peru's economy has come to depend greatly on fish meal as a source of foreign credit.

The American tuna fleet, which in 1969 included 140 ships and employed two thousand men, has come into conflict with Ecuador and Peru. Like the United States government, the tuna skippers recognize only a twelve-mile coastal fishing zone. Fourteen United States tuna ships were seized in 1969 by naval units of Peru and Ecuador for fishing inside the 200-mile limit. One ship, the *San Juan,* was badly damaged by Peruvian machine-gun fire. North American tuna boats would be allowed inside the 200-mile limit if they paid license fees, but the fees are very high and payment would imply acceptance of the CEP claims.

United States legislators have been angered both by loss to our fishermen and by the fact that some of the patrol craft used in seizures by Ecuador have been United States Navy vessels lent to that nation in 1960. Defending the interests of the tuna fishermen, Congress in 1969 passed a law requiring the United States government to pay the fines of ships detained in South America and to reimburse skippers for fishing time lost. The money is to be deducted from foreign aid sent to the nation involved. The law does not seem to have lessened international tension, nor has it reduced the number of tuna boat seizures.

The latest conference between the United States and the CEP nations about the 200-mile limit failed to reach any compromise, and the South American countries will not submit the dispute to arbitration. In Peru, fishing incidents have been overshadowed by conflict about control of land assets owned by United States citizens in that country. Both problems are being considered in diplomatic negotiations. The United States opposes the pushing of fishing rights farther out into the high seas, and therefore comes in conflict with nations that are trying to do so.

Another development in modern fisheries is the use of distant-water fishing fleets, particularly by the Soviet Union and Japan.

The Soviet Union has become one of the world's most important fishing nations, sending fleets to fish in the most distant parts of the ocean. They are seen off the eastern United States in waters traditionally used by coastal fishermen, in the seas off South America, in the Antarctic, and all over the Pacific.

Such fleets are closely coordinated and nearly self-sustaining. A number of catching vessels accompany a factory ship to the fishing grounds. Other ships bring out personnel and supplies or carry fish back to market. Hospital, recreation, and engineering ships are part of the fleet. In a number of vessels, sonar devices that hunt for fish scan large areas of water, so that the factory ship operates efficiently throughout the season. Enormous amounts of fish are harvested, processed, and transported in a short time. Sometimes the efficient distant-water fleets intimidate and anger coastal fishermen still using less advanced fishing methods.

The Soviet fleets are both admired and criticized. Whether Soviet skippers are villains or merely hard-working sea captains depends on one's point of view. A few years ago in Australia there was a public uproar because a Soviet ship had plowed into a group of small Australian boats and scattered both the fishermen and the shrimp they were catching. The press was quick to condemn the Russians. However, later reports noted that, when the Australians explained to the Soviet skipper how he should have approached the shrimp, he readily followed their advice. He also took some of the Australians aboard his ship, offered them caviar and vodka, and gave them medical aid. He is said to have rescued the crew of a local boat that capsized.

Compared with the Australian shrimp boats, the Soviet

vessel was large, modern, and efficient. It had freezing facilities, a fish-meal plant, and equipment for ocean research. But the local fishermen who had resented its presence were mollified by the cooperation of the skipper.

The Soviet Union has effectively turned its fishing fleets into instruments of international good will, and gives leadership and help with fisheries to underdeveloped countries. It has built a modern fish-processing industry for Senegal, given Algeria an ocean-research vessel, and created a fishing harbor in Egypt. In offering assistance, the Soviets try to make their own fishing fleets and those of the receiving country as self-supporting as possible, thus reducing the cost of such aid.

The Japanese are considered by many experts to be the greatest fishing nation in the world today. With a large population and easy access to the sea, they depend heavily on ocean resources for their daily food. As their inshore fisheries have become overused, they have turned to distant-water fleets which appear in every corner of the globe.

At the end of World War II, the Japanese found themselves with old ships, deteriorating equipment, and a severely restricted political situation. Despite these handicaps, they have rapidly returned to a strong position in fishing. Even before the war, they had developed a system of monopolies, dividing their fleets by areas of the ocean and types of catch. The use of licenses now restricts the number of Japanese vessels in any fishery, thus keeping down fleet costs and preventing competition from ruining prices. The government has been active in supplying money for improved ships, for research, and for new technological developments.

While the Soviet Union and Japan have expanded their use of world fisheries, the United States has stood still. The annual United States harvest of fish has changed little since 1945, while the total world catch has more than tripled. Hence the United States has dropped from second to sixth place among the world's fishing nations. The number of commercial fishermen has declined, and the Bureau of Commercial Fisheries acknowledges that the United States fishing fleet is obsolete. Only the shrimp and tuna fleets have done well in the last few years. Fishermen who cling to the independence and satisfaction of their profession are becoming discouraged by low profits. Rightly or wrongly, they blame smaller catches on efficient foreign vessels that work just outside the United States fishing zone. They blame high costs on outmoded maritime laws, some of which date back to 1793.

Most United States fishermen work from medium-sized vessels; only 13,000 boats out of 84,200 are over five tons. They stay close to shore or fish traditional grounds, such as the Grand Banks off Newfoundland, rather than work in distant-water fleets that stay at sea for long periods. In the Pacific, well-organized industries have developed for tuna and halibut and, to a smaller extent, for salmon and crab, and efforts to conserve these species have been reasonably successful. So far, however, United States fishermen have not adopted the use of large, organized fleets.

The greatest hindrance to United States fishing may well be internal regulations that both protect and restrict local fishermen. In the United States, unlike Europe, hunting and fishing have traditionally been open to all, with no private ownership of fish in rivers and streams. The only control over such stock comes from the ownership of river banks, which may be closed to fishermen. At sea, fish belong to any American who catches them. The problem facing lawmakers is to make fishing economically profitable and still be fair to all concerned.

The growth of the United States fishing industry is handicapped by laws requiring that fishing boats of more than five tons be built inside the country. The law originally protected shipbuilders, but it now encourages the use of smaller ships, since these can be bought at lower cost abroad. Imported nets and other equipment are restricted by high import duties. Since labor costs are usually higher in the United States than in Europe and Asia, United States fishermen find it difficult to compete with foreign ships.

The United States has one new vessel that can compare with foreign distant-water ships, the *Seafreeze Atlantic,* launched in 1968. Almost three hundred feet long, with space for nine hundred tons of frozen fish, it has factory equipment in an assembly line that will wash, clean, skin, cut up, freeze, and package fish in a few hours. Fish that cannot be eaten is processed into fish meal and oil. The ship has a speed of over fourteen knots and can travel 26,000 miles from home. The *Seafreeze Atlantic* cost more than five million dollars, half of which was paid by the United States government under its Construction Differential Subsidy Program.

Because of its large initial investment, the ship had to be subsidized by the government, and its fishing will be restricted by the government in order to protect traditional fishermen. Thus its efficiency will be handicapped by complex laws.

In the international field the United States has been successful in making treaties with nations that send distant-water fishing fleets close to her coast. Treaties must be renegotiated every few years because continuous problems arise over conservation of fish stock and interference with fishing gear. Large Soviet trawlers have interfered with local boats along the Atlantic coast and uprooted certain nets in the Pacific. Such matters are settled by horse-trading. When the United States wants the Soviet ships to obey conservation rules, it offers some concession in return. In the Pacific, by an agreement negotiated year after year, the Soviets have given up claims to fish in certain Alaskan waters in return for the right to transfer cargo from one ship to another in protected areas near the Aleutians. In 1969, an Atlantic agreement was signed that prevents Soviet ships from fishing near the continental slope, where fluke and porgie spend the winter. Both species have become scarce and need protection. In return for abstention in this area, the Russians are allowed to transfer fish within the United States fishing zone.

The United States, like Great Britain, continues to favor a narrow territorial sea, and strongly emphasizes the distinction between a fishery zone and a true territorial sea. In 1966 Congress decided to extend the United States fishery zone far enough to provide for conservation efforts in areas that had previously been considered part of the high seas. The Bartlett Act[1] gives the United States an exclusive fishery zone out to twelve miles from the baseline. In this area no other nation may take fish without permission. The twelve-mile fishery zone has enhanced the bargaining position of the United States in relation to other nations whose ships approach its shore.

The United States counts fisheries as a small fraction of the national wealth, and fishing provides jobs for only a relatively small number of men. Food resources are so great on land that, although the United States imports seafood, fish is generally used as a variety food, rather than as a staple item in the daily menu. This has been especially true since the modification of religious rules that previously increased fish consumption. The United States makes little use of ocean products to feed the poor or undernourished population, but processes quantities of second-rate fish for cats.

However, one of the most important new concepts about world fisheries is that they may become a major source of food for expanding populations. Already experts studying overpopulation say that famine conditions will exist in parts of the world in the 1970's, in spite of efforts to control birth rates. But the use of ocean resources to feed masses of people may turn out to be more complicated than it first appears. There is sharp disagreement among experts about the potential food in the sea, and they are not yet able to predict future harvests. Also, people's preferences in food must be considered. In Peru, the coastal population eats quantities of fish, but inland groups in the same nation will not touch it. Food must be in a form that appeals to people, and does not conflict with traditional dietary rules. Ocean resources must be used in some form that can be stored, transported, and cooked easily. Several tons of frozen seafood might go to waste in a tropical country unprepared to handle it. Finally, the cost of ocean products must compare favorably with land resources before they will be widely used.

A new product called *fish protein concentrate* (FPC) is now being made from ocean resources. Although made from whole fish, just as fish meal is, if used for human consumption in the United States, it must be manufactured to Food and Drug Administration standards. A gray, flour-like substance, it can be used with regular flour in making food such as bread or noodles, or it can be mixed into gravies. It has a high protein content and is a more complete form of protein than such vegetables as soybeans, now widely used in place of meat. Because FPC uses the whole fish, including the head, entrails, skin, and bones, some people object to it on esthetic grounds. However, it has no taste or smell of fish. When Danish pastry made with FPC was served at a scientific conference recently, it tasted like any breakfast bun. Nothing indicated that it contained fish concentrate.

FPC has many advantages as an all-purpose additive in the diet. It is cheap because it uses plentiful fish, often those that do not find a ready market in conventional form. The fish need not be cleaned or cut by hand and a processing plant can handle large amounts at one time. FPC can be stored at normal temperatures for long periods and can be transported in bulk. Its main advantage is that it is not a starch, but a highly concentrated, complete protein—a substitute for the meat lacking in many diets.

Fish protein concentrate has been used in small amounts in several parts of the world. Both the United States and Sweden sent FPC to the Nigerian province of Biafra in 1968 to feed starving children. The Swedish relief team kept careful records of the use made of FPC and of milk protein. They concluded that children who were fed FPC recovered faster than those given milk, and developed no undesirable side effects. Apparently some children in Africa react badly to milk protein because they are unaccustomed to this type of food after infancy.

If the protein intake of the world's population could be increased, it might solve problems other than simple hunger. Especially in young children, protein deficiency can cause slow growth and sluggish minds. Even if FPC is not produced in great enough quantities to solve problems of hunger, small amounts of it can make a major improvement in inadequate diets.

New methods of raising fish in enclosures, of feeding them, or of herding them in the sea are being studied in many parts of the world. Aquaculture in fresh or brackish water began many years ago and is now responsible for a significant percentage of the world's total fish catch. Carp, milkfish, mullet, rainbow trout, and shrimp are raised in artificial ponds, either with or without supplementary food. In the United States, cultured catfish from Mississippi, Arkansas, and Louisiana produced, in 1968, a harvest worth 4.6 million dollars.

True farming of the sea is not widely developed, except for oyster culture which dates from ancient times. The Japanese are leaders in experiments with hatching and rearing fish, often in net enclosures in the sea. The University of Rhode Island is sponsoring an experimental fish-farming project which will raise rainbow trout, salmon, bluefish, and striped bass. Steel pools with vinyl liners will be filled with piped sea water. The water will be aerated and the effects of pollution studied. The project will try to develop cheap, nutritious diets for the fish. The head of the project, Dr. Saul Salia, believes that luxury fish such as trout and salmon can be produced more cheaply through aquaculture than through commercial fishing.

Many theories have been advanced about herding fish in the open sea, although few have been put into practice. Perhaps fish can be driven like cattle and enclosed by fences made of air bubbles, which fish hesitate to cross. Since fish thrive in areas where upwelling currents bring nutrients from the bottom into sunlit upper waters, upwellings might be produced artificially, perhaps using nuclear power. Fishermen have already tried creating artificial reefs to attract fish and lobsters, using everything from rubble to old automobiles.

In sea water, old automobiles disintegrate after three or four years (which is bad or good, depending on whether one is trying to attract fish or dispose of automobiles).

When fish are raised in enclosures, or even in nets or fences, their ownership is clear. Herding fish in the sea, improving their habitations, or increasing their food supply raises questions about fishing rights. No one is likely to invest in underwater concrete habitats to attract lobsters if the harvest is open to all. On the high seas, it would be difficult to claim ownership of a far-roaming species of fish that had grown up in a protected and controlled nursery. The old Roman law that fish are wild animals by nature will have to be modified if modern technology perfects methods of raising fish like domestic cattle.

As population increases and hunger becomes more prevalent, fish and ocean resources are being given more attention than ever before. International law stresses conservation measures based on sound scientific knowledge. Many nations are claiming greater areas along their coasts for fishing, and spending money on distant-water fleets. They are developing new ways of increasing fish stocks, of harvesting fish and processing food from the sea. A legal structure has been built up, resting on commissions, conventions, treaties, judicial decisions, and international conferences. However, the last codification of the law was at Geneva in 1958. Since that time, new developments have been dealt with chiefly in treaties or regional conferences. If it is to remain valid, the Law of the Sea must keep up with new technology in fishing, with conservation needs, and with increased claims of ocean ownership.

Ocean Resources: Animals

26 Fur seal being captured on ice

27 Tying up finback whale off Eureka, California

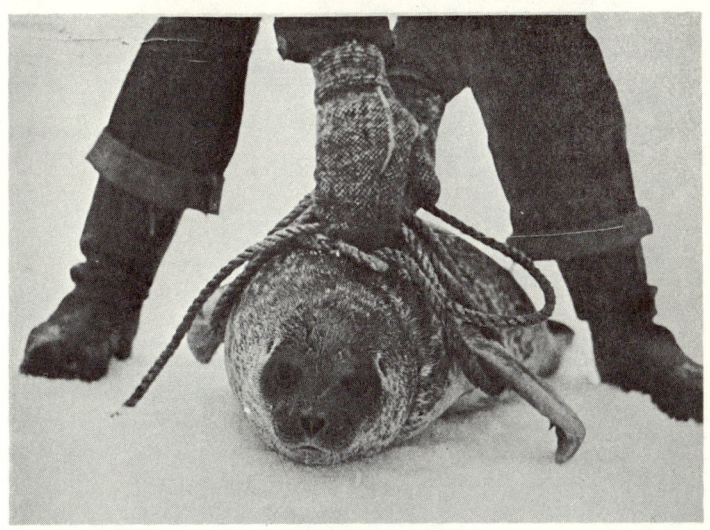

28 Three Adelie penguins "hold a conference" near the *U.S.S. Bear* at the Bay of Whales

29 Walrus on ice

30 Hauling a catch of codfish aboard the *Gloucester*, off New Brunswick coast

31 A huge, funnel-shaped net, buoyed by floats, being set from a fishing dragger

32 Part of the vast fleet of Russian and Polish fishing vessels in the English Channel, chasing seasonal swarms of herring

33 Russian fishing trawler detained by the authorities for illegally fishing inside Canadian waters

34 Peruvian patrol boat, the crew of which boarded the American fishing boat *Marina* off San Diego, California, and towed her 500 miles to the port of Talara, Peru

35 A Soviet "mother" ship with several trawlers rides at anchor at Norfolk, Va. as an American ship pulls alongside for a closer look

CHAPTER 6

The Technological Revolution: Submersibles

The deepest known place in the world lies at the bottom of the Pacific Ocean in the Mariana Trench, a canyon that drops far below the normal level of the sea floor. This trench lies southwest of Guam and contains an area called the Challenger Deep which—over 35,000 feet beneath the surface, seven miles down—is farther below sea level than any other accessible place on earth. In 1960, Jacques Piccard and Donald Walsh descended in the bathyscaphe *Trieste* to this awesome depth and returned successfully, becoming the first and only men to reach this deepest part of the ocean.

The *Trieste* was built in 1953 by Auguste Piccard and his son Jacques after the original bathyscaphe, *FNRS II* had been sold to the French Navy. Like its predecessor, the *Trieste* was made up of two parts: a large, balloon-like body filled with gasoline, which is lighter than water, and a spherical cabin hung beneath it. Iron shot held by electromagnets was added to increase the total weight of the boat and enable it to sink. It ascended when the iron shot was dropped and the gasoline-filled float rose in the water, just as a balloon rises in the air.

A number of improvements were made in the design of the *Trieste*. In the *FNRS II,* the two passengers entered the steel cabin before the bathyscaphe was lifted off the surface ship and filled with gasoline. Returning to the surface, the men reversed this process, remaining in their cramped quarters until the boat was emptied and lifted from the water. The cabin of the *Trieste* was reached by a ladder inside a special compartment running down through the gasoline float. Since this access compartment was flooded during a dive, it did not have to withstand great pressure. After a dive the water could be forced out of it by compressed air. Other technical improvements made the *Trieste* more seaworthy and effective than the earlier model.

Jacques Piccard made a number of dives with the *Trieste* in the Mediterranean during the 1950's, setting new depth records and gathering scientific information. After such trials had proved its abilities, the United States Navy bought the bathyscaphe. Piccard accompanied the vessel as a consultant.

Before a dive to the deepest part of the ocean was attempted, Piccard designed a new cabin of greater strength, which was built in Germany. Still approximately of the same size — seven feet in diameter — the new sphere was heavier and had a skin five inches thick. Inside this small sphere, the walls were covered with instruments for controlling the ship and operating equipment. Viewports made of plexiglass allowed visibility. Lights, pressure gauges, and various other scientific instruments were outside the cabin. Because of difficulties in forging, the new cabin was built in three sections, bonded together with epoxy resin. In one of the dives preparatory to the attempt at the Challenger Deep, the epoxy-resin seal failed, causing a loud explosive noise, which was frightening but not serious. Fortunately it was possible to repair the seal and continue the diving schedule. (Under pressure, especially in the deep sea, parts of a sphere are automatically held together by the forces pressing in on them. The deeper a sphere goes, the tighter its sections are forced together. For this reason, many deep-ocean vessels are in the shape of one or more spheres, or of a cylinder with spherical ends.)

Originally, three dives were planned for the Challenger Deep, but, as time passed, bad weather in that part of the Pacific made only one attempt possible. Donald Walsh, the officer in charge of the project for the United States Navy, was scheduled for this single dive, along with another navy man. However, Piccard, showing his contract to prove that he was to be present on any dives that "present special problems," became the man who finally was chosen to accompany Walsh.

Despite very heavy seas, the attempt on the Challenger Deep began on January 23, 1960, at eight A.M. By letting out either a little gasoline or a little iron shot, the men could regulate the downward speed of the dive, but they had only limited control over the movement of the bathyscaphe in other directions. Contact with the sides of the Mariana Trench or other obstacles might have proved serious.

One unexpected event did occur. At 32,000 feet, a strong muffled explosion shook the sphere as if it were caught in a small earthquake. A plastic panel in the access compartment of the gasoline float had cracked. However, the cracked panel remained tight enough to enable the men to empty the compartment when the dive was completed.

A little after 1:00 P.M., the bathyscaphe touched bottom. For twenty minutes, Piccard and Walsh made scientific observations. Through the viewports they saw a flat, flounder-like fish, the presence of which proved that complex life does exist in the deepest part of the ocean. By this time, the temperature inside the sphere had dropped to 50 degrees Fahrenheit, chilling the men, who had little space in which to move. For lunch they had brought only chocolate bars, which they decided to save in case they were delayed in leaving the sphere. After the iron shot had been released, the bathyscaphe began its slow ascent to the surface, where it arrived at five o'clock. Piccard and Walsh, after nine hours in the cabin, climbed through the entrance compartment and were greeted enthusiastically by newsmen and sailors aboard the surface ship.

The descent of the *Trieste* into the Challenger Deep marks the end of an era. The depth record established is not likely to be broken. Except for a slightly deeper spot found nearby by the Russians in 1957, the *Trieste* touched bottom at the deepest place known to man. The *Trieste's* dive was a one-time attempt, unlikely to be repeated, for it was an experimental feat, important in theory but of no immediate purpose. At that time, fleets of military submarines that were large, expensive, and deadly were built in quantity from designs evolved over the past fifty years. The few vessels like the *Trieste* were tentative experiments, entirely new attempts to build ships to help men live and work in the ocean.

In less than the ten years following 1960, a tremendous revolution in technology has occurred, particularly in the construction and use of scientific and industrial submersibles. If big military submarines are like jet airliners, these new underwater vehicles, called both submersibles and mini-subs, are the helicopters of the deep. Very little new scientific theory has gone into their development, but, in an astonishingly short time, new technology has been worked out to make them practical and safe.

Compared with military submarines, these new underwater vehicles are small—carrying from two to six men. They are built to maneuver on the bottom rather than to fight or travel considerable distances beneath the surface. Because most submersibles can travel only a short distance under their own power, they are transported on surface ships, or towed behind them, to their work areas. Few can move with any great speed, but all are designed to be maneuverable in six directions if possible: forward and back, right and left, up and down. They have viewports through which the operators can see the water around them, and external lights, cameras, instruments, and gauges to help them make contact with the ocean world. Many of the vessels have mechanical arms or various forms of scoops or hooks manipulated from inside the cabin. Some have lock-out chambers to allow divers to leave the vehicle in deep water. In addition to their usefulness in scientific studies, the new submersibles are being built to help men carry out a variety of tasks: finding lost objects, rescuing men from downed submarines, helping divers living in underwater habitats, inspecting cables, and working on offshore oil wells.

Only three years after the dive to the Challenger Deep,

the *Trieste* was used to locate the wreck of the nuclear submarine *Thresher,* lost on a test dive 315 miles east of Boston. After surface vessels had scanned the ocean floor with underwater film cameras, television, magnetic detectors, and sonar to find the general location of the wreck, navy operators took the *Trieste* down more than eight thousand feet on a number of dives. The first evidence they saw on the bottom was a plastic shoe cover of a type used on nuclear submarines. Later they were able to distinguish a mass of twisted metal, including battery plates, shredded cables, and a section of the superstructure. Although the *Trieste* did not conclusively establish the cause of the *Thresher* disaster, it was successful in finding the remains of the ship and describing its appearance. The bathyscaphe is the only vessel the United States Navy possesses that could go to the necessary depth in the ocean.

Submersibles were used in another undersea search in 1966 after two United States planes collided in the air over Palomares, Spain. One of the planes carried four untriggered hydrogen bombs. The first plane, a SAC B52 bomber on routine patrol, was approaching a tanker plane for mid-air refueling when 40,000 gallons of jet fuel in the tanker exploded into flames. The aircraft disintegrated. Only four of the eleven crewmen in the two planes managed to parachute to safety.

Although carried by parachutes with automatic releases, three hydrogen bombs broke up on land, and the pieces had to be retrieved from the Spanish countryside with great care. Public opinion was aroused by the possible danger from scattered nuclear material. When it became apparent that the fourth bomb had fallen into the sea, the United States made every effort to recover it.

The search began with Navy frogmen who swam shoulder to shoulder to a depth of 130 feet. There hard-hat divers took over and continued down to 400 feet. In this shallow water, tiny two-man submarines, called Cubmarines, were also used, but without results. Beyond that depth the Navy marked off the area in grids on a map and used all the surface-scanning methods available, but still failed to locate the bomb. A local fisherman pointed out a spot some miles offshore where he said he had seen the bomb fall, but because many local stories were circulating, his word was not at first taken seriously. Later, to search this deeper area, the Navy called in two deep submersibles available at the time.

The larger of the two vehicles, the *Aluminaut,* was brought from the United States on a surface ship. Fifty feet long and constructed of eleven cylindrical aluminum rings bolted together, the *Aluminaut* looked like a small conventional submarine. Able to carry three men to a depth of 15,000 feet, it had a cruising speed of three or four knots and could remain submerged for thirty-two hours. Its lack of maneuverability made observation of the sea floor difficult, but it could mark a given location by remaining in one place on the bottom for some time.

The second submersible, the *Alvin,* was small enough to be brought to Spain on an Army transport plane but had to be reassembled on arrival. Short and stubby, the *Alvin* had a pressurized, seven-foot sphere for a cabin, and an external hull twenty-two feet long. It usually carried two men, a third on a surface ship directing operations. Liquid mercury, pumped back and forth between various tanks, controlled the vehicle's trim just as water controls the trim of conventional submarines. Able to travel at only two knots and to stay submerged for eight hours, the *Alvin* still had a definite advantage in its ability to move in all directions—to "squiggle"—so that men looking out from the viewports could scan the ocean bottom. A mechanical arm later attached to the front of the vehicle proved very useful, although difficult to control.

Even with the information from the local fisherman, two weeks passed before the *Alvin,* searching systematically across the Navy's grid, found a skid mark where the bomb had fallen and then slid in the mud. At this point, five miles off Spain in 2,500 feet of water, the bottom was a long slope without vegetation, covered with muddy sediment that rose in clouds whenever the *Alvin* stirred it up. Under these conditions, the searchers lost sight of the skid mark before they could follow it to the bomb. Twelve days were spent finding the skid mark again. Finally, the bomb was located, resting precariously on the slope, entangled in its parachute. While the *Alvin* returned to the surface to fit on its mechanical arm, the *Aluminaut* remained on station marking the bomb's location.

Bringing the bomb to the surface proved that the submersibles, though useful in the search, had definite limitations. After the *Alvin* had gotten a line on the bomb, barely avoiding the entangled parachute, the bomb broke away. Another nine days passed before it was relocated. Then, to bring up the bomb, the Navy used an unmanned machine, called *CURV,* which attached grapnels to the parachute.

One legal point was raised by the location of the submerged bomb, five miles off the Spanish coast and therefore on the high seas. During the search, Spanish authorities closed off the area, and the United States Navy sent out two destroyers to warn off a Soviet trawler that tried to approach. Fortunately, the Soviets never questioned this action, and the world press did not ask by what right the United States could prohibit anyone from entering an area of the high seas. Perhaps everyone was more concerned with the danger of possible nuclear contamination than with the freedom of the seas.

The Spanish people, especially those whose farms were inspected for radioactive material, were frightened by the bomb, and did not realize that nuclear weapons cannot cause an atomic explosion unless they are activated. The United States, affected by the unfavorable publicity, tried to reassure everyone and to clean up the debris as rapidly as possible. The *Alvin* and the *Aluminaut* were vital factors in the program to overcome the effects of the disaster.

Under normal conditions, the *Alvin* is operated as a research vehicle by the Woods Hole Oceanographic Institution. In 1967, when it was being used to study manganese and phosphate deposits in the water off Georgia, it had the strange experience of being attacked by a fish. As one of the *Alvin's* crew watched through a viewport, a swordfish rose suddenly from the bottom, and without hesitation, rammed the submersible. Its sword missed a plexiglass viewport and became wedged in a joint of the exterior hull. When the *Alvin* surfaced, the crew spent two hours removing the sword, which had done little damage. After the fish had been inspected, measured, and weighed, it was cut into steaks and eaten for dinner.

The *Alvin* was lost in 1968 on a mission to explore a canyon in the ocean floor 120 miles south of Cape Cod. When not in the water, the submersible was carried on a cradle slung between the two hulls of its mother ship, the catamaran *Lulu.* In preparation for a dive, with two men inside and a third man about to enter, the *Alvin* slid into the water because one of the forward cables holding the cradle broke loose. The three men were able to scramble to safety, but they were unable to close the submersible's hatch. Water poured into the vehicle, which broke completely free, sinking in 5,000 feet of water.

After being given up for lost and lying on the bottom for nearly a year, the *Alvin* was salvaged by the joint efforts of a Navy surface ship and the *Aluminaut.* With its manipulator arm, the *Aluminaut* lowered a toggle bar, attached to a nylon line, into the still open hatch of the small submersible. The Navy ship raised it to a depth of eighty-five feet, where divers slung a net under it and fastened on salvage pontoons. Towed to calm waters, the *Alvin* was pumped out and taken by barge to Woods Hole.

Probably the most famous of the small submersibles is Jacques-Yves Cousteau's *Diving Saucer (Soucoupe Plongeante),* used in his oceanographic films and his experiments in underwater living, Conshelf II and Conshelf III. A pioneering effort in this type of vehicle, the idea for the *Saucer* began as early as 1955, when Cousteau, working with free divers, realized the need for a vehicle to help men explore the sea at greater depths, for longer periods of time, and with greater safety and comfort. Such a machine would also enable scientists to observe the ocean floor, take photographs, and gather samples.

Diving Saucer II (the first model was lost because a cable broke during an early tethered unmanned dive) was completed in 1959. Its pressure hull, shaped like a thickened disk, six and a half feet in diameter, allows two men to lie

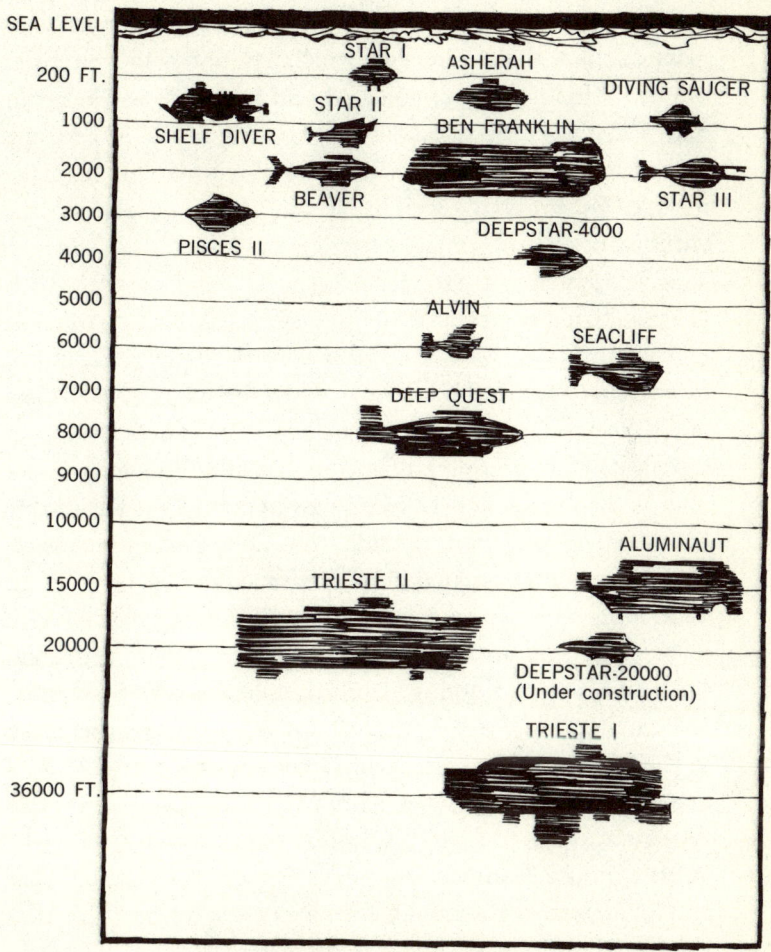

TYPICAL SUBMERSIBLES
SHOWING THEIR MAXIMUM DEPTHS

Deepstar-2000 (not shown) is now completed, but was undergoing sea trials at the time this diagram was prepared

prone, looking through the viewports, or to sit upright in a cramped position. Outside the hull and enclosed by a fiberglass shield is the equipment, which can withstand pressure: motor, altitude control, and lead-acid batteries suspended in oil. With a set of cast-iron weights the *Saucer* is adjusted to neutral buoyancy and is then weighted enough to make it sink. By dropping weight, it can return to the surface without depending on its batteries for power.

A jet system of propulsion can be handled with such delicacy that the operator can adjust the *Saucer's* movement in any direction. Motion is also controlled by 275 pounds of mercury pumped back and forth between two ballast tanks. Forward speed is less than one knot. The life-support system consists of a tank of medical breathing oxygen under one of the bunks and trays of a compound called baralyme to absorb the carbon dioxide. As Cousteau's underwater films have demonstrated, the *Diving Saucer* is adapted to many uses and can be handled skillfully. It is small enough to be carried on an aircraft, yet strong enough to withstand the pressure at a working depth of 1,000 feet.

From 1960 to 1964 the *Saucer* was used for scientific dives in the Mediterranean, after which it was leased to the United States and used in a series of dives off the California coast. At about the same time, Cousteau was working with the Westinghouse Company designing a new submersible that employs some of the design features of the *Saucer*. This new vehicle, the *Deepstar-4000,* has been used by the Navy and other contractors for research. Part of a family of new submersibles, the *Deepstar-4000* has been followed by the *Deepstar-2000,* which will be used to explore the Cobb Seamount, an underwater mountain off the northwestern coast of the United States.

Many new submersibles have been developed recently. Some of them, as small as the Cubmarines, cost less than 50,000 dollars, but others, like the *Aluminaut,* require a major investment to design and build. Private industry has built the majority of them. Most are designed to meet specific needs in science and industry.

One of the earliest submersibles, the *Asherah,* was constructed in 1964 by General Dynamics for the University of Pennsylvania. Able to hover and maneuver easily, the two-man, sixteen-foot vehicle suits the needs of archaeologists investigating ancient wrecks under water. After it was used to explore a Byzantine hull in the Aegean Sea, it was sent to Hawaii, where the Bureau of Commercial Fisheries employed it to study the reactions of fish to various kinds of bait and tackle. The *Asherah* was followed by a series of vessels from the same company, including the *Star II* and *Star III.* They have been used for undersea research and for commercial applications such as underwater cable inspection.

Perry, the company that developed the Cubmarines, has worked with Edwin Link to produce a larger version of these vehicles, called the *Shelf Diver.* Twenty-two feet long, it has two separate compartments, a chamber at sea-level pressure for the crew and a high-pressure lock-out chamber from which divers can exit directly into deep water. Vehicles with lock-out chambers promise to be useful to the oil industry for work on underwater installations.

The *Deep Quest,* Lockheed's thirty-nine-foot submersible, is a larger vehicle, in a class with the *Aluminaut.* It can carry three or four men, travel at two knots for twenty-four hours, and work at an 8,000-foot depth. On a practice mission in October 1969, the *Deep Quest* became entangled in a plastic line that was attached to a large test object it was trying to raise from the bottom. When the *Deep Quest's* plight became known, several other submersibles were made ready for the rescue. A small vehicle, the *Nekton*, was chosen. It dived, sawed through the line with a knife blade held in its manipulator, and freed the *Deep Quest*.

Another boat designed for industry is North American Rockwell's *Beaver*, a medium-size submersible that can work with divers at 2,000 feet. Learning from the experience of the *Alvin*, which used one mechanical arm with difficulty at Palomares, the designers of the *Beaver* have supplied two external arms that can perform such operations as threading a cable through a metal eye. The boat was built specifically for the oil industry and is to be used on pipelines and underwater wellheads.

The United States Navy has a comprehensive program for underwater vehicles, each created to fill a definite military need. After the loss of the nuclear submarine *Thresher*, in 1963, the Navy realized the need for a submersible capable of reaching a sunken submarine and bringing off any survivors. The Deep Submergence Rescue Vehicle (DSRV) has a pressure hull of three connected spheres, plus a skirt, or mating bell, that hangs down to fit over the hatch of a disabled submarine. Although the Navy expects to build enough of these submersibles to reach any spot in the ocean within twenty-four hours, so far only one has been launched—in 1970. When not on emergency duty, it can also do scientific tasks.

The Navy's newest submersible, the *NR I*, is 112 feet long, almost big enough to be a conventional submarine. It runs on nuclear power. Work is in progress on a Deep Submergence Search Vehicle (DSSV) that will locate objects down to 20,000 feet.

Other nations, besides the United States and France, have submersible programs, some of which are quite extensive. As early as 1951 the Japanese were using the *Kuroshio*, a tethered, bell-shaped vehicle that could be rotated but had no forward propulsion system. It was used to study fish distribution and the scallop population, matters of great importance to the Japanese, who get 64 percent of their total protein food from the sea. A newer version, the *Kuroshio II*, moves freely and has a motor, although electric power is still supplied from the surface. A more conventional vehicle, the *Yomiuri*, launched in 1964, has a cylindrical pressure hull with the usual lead-acid batteries, external motors, viewports, and a small claw-like manipulator. Like a military submarine, it has diesel engines that propel it on the surface and can recharge the batteries. The *Yomiuri* has made biological and geological surveys of the continental shelf off Japan and has conducted seismic studies after an earthquake.

The land-locked Swiss have built one submersible, the *Auguste Piccard*, designed and built in 1963 by Jacques Piccard and launched in Lake Geneva. The cylindrical hull, large enough to hold forty passengers, has comfortable seats and individual viewports, like an airliner. During the World's Fair at Geneva, the Swiss used the *Auguste Piccard* as a tourist attraction, taking load after load of passengers down to view the bottom of the lake. It successfully carried 33,000 passengers and completed 1,300 dives. The ship has been purchased from the Swiss by a Chicago company that will use it with a giant oil-storage tank submerged in 158 feet of water in the Arabian Gulf.

A Canadian salvage company is using a series of submersibles, the first of which, *Pisces I*, brought up a ninety-ton tug from 670 feet of water in Howe Sound, north of Vancouver. In one of the deepest major salvage jobs ever undertaken, *Pisces I* cleared debris and wreckage from the sunken boat, rigged lifting cables, and guided the raising operation. The Canadian government wanted the tug in order to discover what had caused its loss. In ten years, twenty-three similar vessels have been lost in the same waters, and the Canadians have been searching for some explanation that will prevent further disaster.

One new United States submersible, the *Ben Franklin,* was developed as a joint project by Jacques Piccard in Switzerland and the Grumman Aerospace Corporation in the United States. Like the *Auguste Piccard,* it was built near the Swiss Alps; then it was brought by rail and ship to Florida, where it was first tested in the water.

The *Ben Franklin* is a 130-ton vessel capable of operating at 2,000 feet and carrying six men for more than a month. Because it has twenty-five tons of lead-acid batteries in its keel, it provides power for extended scientific missions. Unlike many of the smaller submersibles, it has a life-support system that allows it to operate without coming to the surface every few hours. It has a fifty-foot cylindrical hull with hemispherical ends and twenty-nine viewports for observation. Made of three-inch thick plexiglass, the viewports are shaped like truncated cones—twelve inches in diameter on the outside and narrowing down to five inches on the inside. Because of this shape, water pressure outside constantly forces them into their steel sockets, automatically preventing leaks.

Four motor-driven propellers are located at the four corners of the boat and can be controlled in a variety of modes to maneuver the vehicle. Ballast tanks can be flooded or

EXPLORING OCEAN FRONTIERS

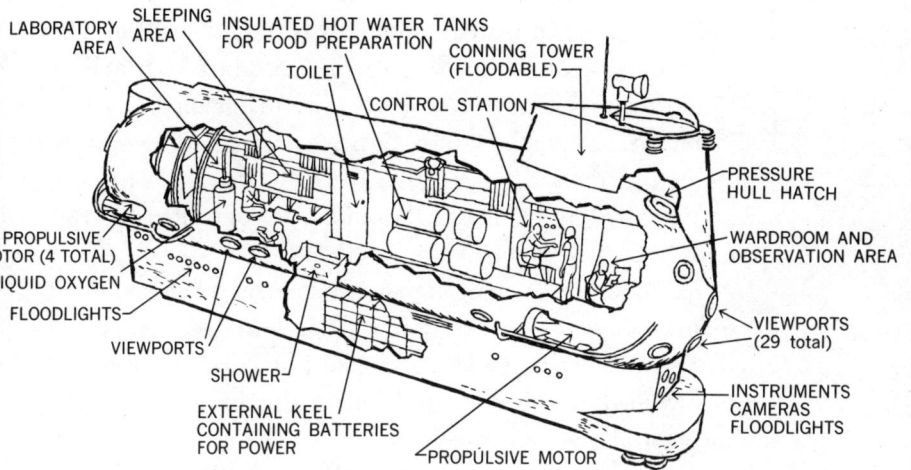

BEN FRANKLIN—2000 FT. DEPTH SUBMERSIBLE

emptied with compressed air to control depth, and 10,000 pounds of iron shot can be dropped to bring the *Ben Franklin* to the surface in an emergency. There is a hatch at each end of the boat and a conning tower that rises above the water for surface travel. During a dive the conning tower is flooded. The heavy batteries in the keel help to stabilize the boat.

Military submarines, with their large power plants and great overall weight, easily support a crew during long voyages. The *Ben Franklin* is one of the few privately owned submersibles that can sustain men for a month under water. The life-support system includes atmospheric controls, food and water supplies, and waste disposal. The atmosphere is at sea-level pressure and has the same chemical composition as air on land. Oxygen consumed by the crew is replenished from two large, insulated containers of liquid oxygen. The carbon dioxide of the men's breath is absorbed by panels of lithium hydroxide open to the air; they maintain the carbon dioxide at acceptable levels. Other contaminant gases are absorbed as the air is circulated through activated charcoal, like that used in gas masks. Humidity is controlled by bags of absorbent silica gel.

The food, on a long mission, consists of freeze-dried products mixed with hot water from four large, super-insulated tanks. These tanks, developed from space technology, can be filled at the dock and will maintain a temperature above 150 degrees for over a month. Cold fresh water in the *Ben Franklin* is stored in other tanks within the hull. After water has been used in personal washing, it is stored in a special waste tank and used again in the toilet. The toilet room has a specially designed fan and air-collecting system to prevent odors from contaminating the atmosphere. Waste products are chemically treated and collected in tanks in the bottom of the hull.

The *Ben Franklin's* life-support system is completely closed. Nothing enters or leaves the submersible until the end of the mission. This system has the advantage of maintaining the submersible at a constant weight, and it avoids pipes running in and out of the pressure hull. The only opening to the outside is a small lock-out chamber big enough to hold film, records, or other messages for the surface. With two pressure-tight doors, it can be closed inside the submersible before its outer door is opened in the water.

Inside, the *Ben Franklin* looks like a yacht, with upper

and lower bunks, a galley, and living space. The forward compartment is a combined messroom, wardroom, and observation area. With its five viewports, it resembles an inside-out aquarium, for fish often swim up to the ports to look in. The aft compartment is similar but smaller. Near the middle of the boat are the toilet and shower rooms. Behind them are the six bunks, work tables, and space for the special instruments required on each mission.

Aft of the forward compartment, the galley is on one side, and the control station opposite it. Here the pilot has controls for the motors and valves for blowing or flooding the ballast tanks. Because most of the submersible's controls run on electricity, there are switches, warning lights, and a bank of electrical indicators. The pilot, who can tell his direction from an aircraft compass, steers either by using an electrically powered rudder or by differentially varying the thrust of the port and starboard motors. He determines his depth and the distance to the bottom by means of water-pressure gauges and a sound fathometer. His communication with the surface is provided by an underwater telephone that transmits on sound frequencies instead of on radio waves, which will not penetrate water. Instead of radar, which also will not function under water, the pilot has a scanning sonar that presents a radar-like picture but has considerably less range. Underwater sound is effective only for a few thousand feet. Underwater television, mounted on the bow of the boat, can be aimed by remote control to give the pilot a picture of what lies ahead. The submersible is usually controlled by one man from this single control console.

During the summer of 1969 the *Ben Franklin* completed its first project, the Gulf Stream Drift Mission. With Piccard as scientific leader and five other men to operate the boat and make special studies, the *Ben Franklin* entered the Gulf Stream off the coast of Florida and, letting the current carry it along, surfaced a month later near Nova Scotia. It carried cameras and scientific instruments outside the hull, and every

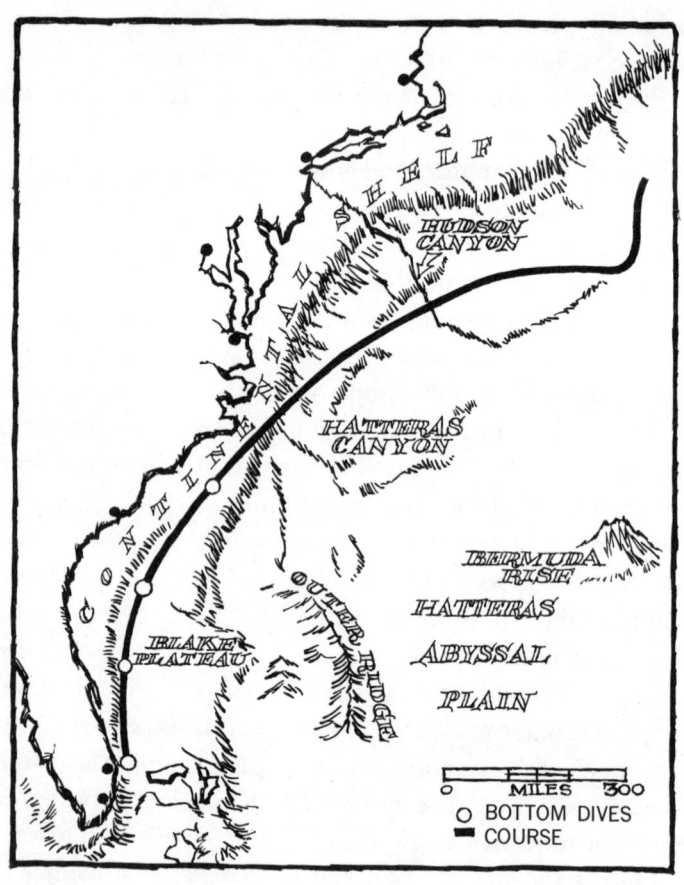

ROUTE OF BEN FRANKLIN

available space inside was filled with recording instruments.

A number of government agencies participated in the Gulf Stream Drift Mission. The Naval Oceanographic Office supplied one of the crew members, and the space agency, NASA, set up a program to monitor aspects of the trip that resembled space conditions. The *Ben Franklin,* on a month-long mission, offered an excellent testing ground for space problems. Its closed atmosphere was similar to that of a space station, and the isolated environment was used to test physiological and psychological effects on the crew and the possible development of new bacterial strains under closed conditions. For this purpose, a mobile submersible has an advantage over a closed chamber on land, for the crew is performing genuine scientific tasks. The stress and isolation under hazardous conditions duplicate situations that might occur in space.

Although there were similarities between the Drift Mission and the flight to the moon, there were major differences. The *Ben Franklin* was carrying out one scientific mission among many under the surface of the ocean, and it had several alternate ways of returning to the safety of the surface if danger threatened; and the Drift Mission, including the total cost of designing and building the *Ben Franklin,* cost less than one day of the ten-year Apollo space program.

To the disappointment of sensation-seekers, the voyage was rather uneventful. No sea monsters were sighted, although the submersible was attacked by a broadbill swordfish while on an excursion to the bottom. Off the Carolina coast, the *Ben Franklin* encountered internal waves in the Gulf Stream at a depth of 600 feet, with a vertical motion of about 150 feet every six minutes. The ocean floor off the Carolinas was found to be particularly rough and hazardous, with steep hills over 100 feet high. Most of the time the submersible cruised stably at a depth between 500 and 700 feet, using no propulsion. Six trips to the bottom, ranging to 1,800 feet, were made between Florida and Cape Hatteras for bottom surveys.

At one point, south of Cape Hatteras, the *Ben Franklin* was forced out of the main axis of the Gulf Stream by a powerful eddy to the west. After five hours of using power to get back into the Gulf Stream, the vessel was forced to surface and allow its surface vessel, the *Privateer,* to tow it fifty miles eastward to get back on course. The hatches were kept closed during the tow in order to maintain the sealed condition of the boat.

The crew emerged from the *Ben Franklin* healthy, in good spirits, and still speaking to each other. Because of the haste in getting the mission underway before the hurricane season, the crew had had little time to train together, and yet they succeeded in becoming a unified group to accomplish their mission. Lessons learned in such an underwater project may be valuable in planning space missions of the future.

The Gulf Stream Drift program proved that a submersible can carry all the instruments normally installed on large surface oceanographic vessels and, at the same time, avoid the wind and the waves that affect surface ships. It can provide comfortable living quarters for the crew and enough power to operate cameras and instruments. Best of all, the *Ben Franklin* gave its scientists a close-up view of underseas phenomena, a window looking out on the world beneath the ocean.

Bomb Hunt

36 Wreckage of B52 bomber on beach near Palomares, Spain. The U. S. Navy ship in the background is participating in the search for the missing nuclear bombs

37 American technicians, wearing protective clothing, go over a field near Palomares in the continuing search for the bomb

38 The Cubmarine is lowered into the water from the guided missile cruiser *U.S.S. Boston* to continue the search

39 Technicians giving a final check to *Aluminaut* before it makes the 2,500-foot dive to search for the bomb

40 *The Alvin*, which located the bomb

41 CURV, which lifted the bomb to the surface

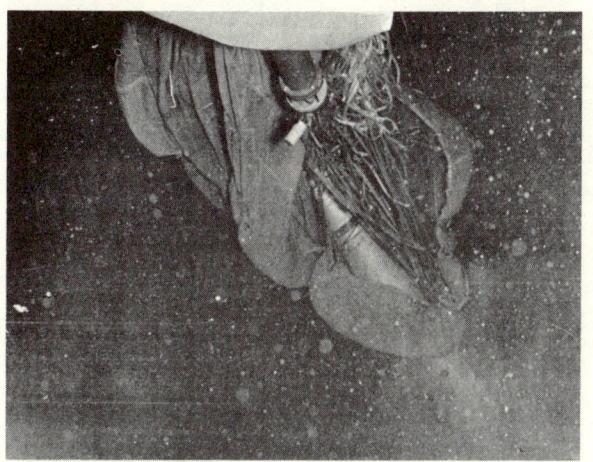

42 The bomb underwater in the process of recovery

43 After recovery, the bomb, still partially wrapped in the shrouds of its parachute, being lashed down on board the submarine *Petrel*

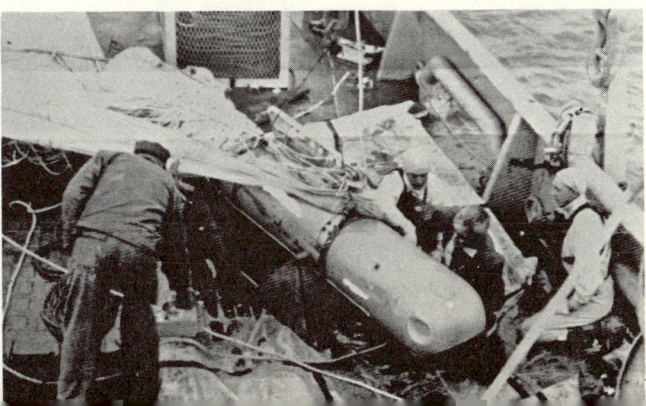

44 Standing behind the recovered bomb, Admiral William S. Guest (second from right), Commander of U.S Navy Task Force 65 is being congratulated by General Arturo Montel, Spanish Coordinator of Recovery Operations

45 Hundreds of barrels of soil from farms near Palomares are being gathered for shipment to the United States, where the soil was to be examined for radioactivity

Ben Franklin

46 Erwin Aebersold, Swiss crewman of the *Ben Franklin* (in the rear) checks the food and water rations for the expedition

47 Two cranes lower the *Ben Franklin* into the water at Palm Beach, Fla.

48 *Ben Franklin* under water off Palm Beach Inlet

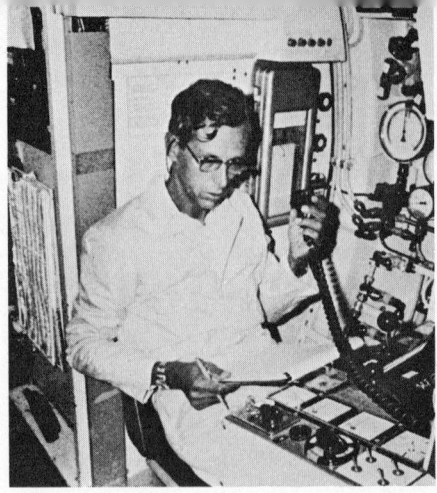

49 Jacques Piccard communicates with the surface

50 Jacques Piccard and five companions surfaced (after a month underwater) about 300 miles south of Nova Scotia

51 Four of the six crewmen being transferred to U. S. Coast Guard cutter after surfacing. Jacques Piccard is in the foreground on the right

Habitats

52 Cousteau's Diving Saucer II about to be transported by air from Marseilles to San Diego, where it was used for submarine experiments

53 Large booms and many cable lines help in one of the final steps of installation of another of Cousteau's habitats

54 Launching, in 1965, off Monte Carlo of one of Cousteau's habitats

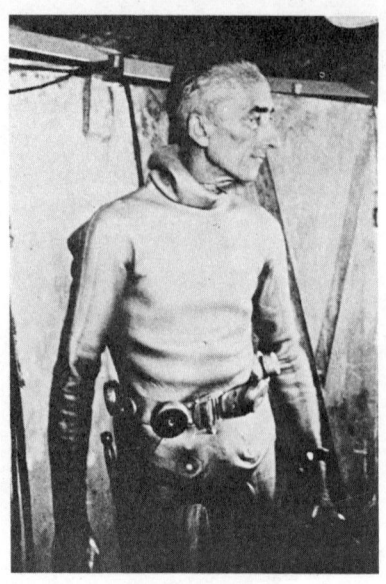

55 Jacques Yves Cousteau in diving gear in the film "World Without Sun"

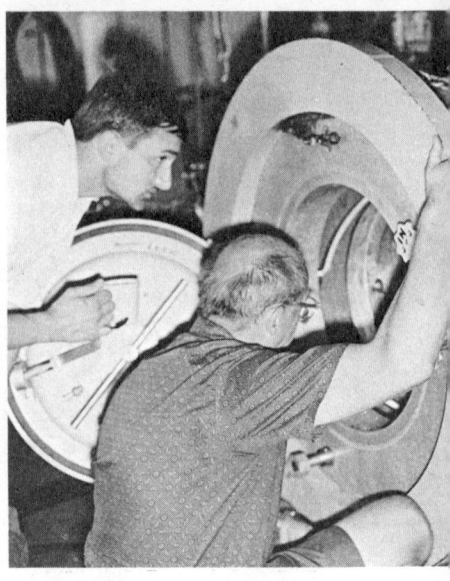

56 Robert Stenuit with Edwin Link just before Stenuit descended in Link's cylinder off Villefranche, French Riviera

57 Jon Lindbergh with Edwin Link, looking over Link's SPID (submerged, portable, inflatable dwelling)

CHAPTER 7

The Technological Revolution: Habitats

In Jules Verne's *Twenty Thousand Leagues Under the Sea,* men living in an underwater structure put on diving suits and went out through air locks to harvest sea plants and herd schools of edible fish. Ignoring many of the real difficulties of living under water, Verne, in conceiving his vision, assumed that, given adequate air to breathe and suits to keep them dry, men could flourish on the ocean floor regardless of depth and pressure. Factors such as cold and lack of visibility were overcome as a matter of course.

As reality has overtaken the dreams of underwater living, practical problems have had to be overcome one by one. After World War II, Cousteau and others developed effective ways for swimmers to work freely under a few feet of water. Hard-hat divers had already been down to greater depths; and, like caisson workers, they had discovered the problems that come from deep-sea pressure. By 1960, scientists had worked out theories of how human beings could live in various depths of water and overcome the difficulties involved. In 1962, the first man stayed down at 200 feet for twenty-four hours. Then the rush began. In the years between 1962 and 1968, more progress was made in teaching people to live under water than was made in the fifty years before that time. It is a difficult, dangerous, complicated process, quite unlike the simple world of Captain Nemo, where the only enemies were big fish and scheming men. Once men seriously decided to try living under the ocean, they needed money, engineering technology, and great courage to make their dream a reality.

There is little value in trying to determine which man-in-the-sea project first reached each landmark in underwater living. Official records may someday establish that one program accomplished a specific feat before another, but ocean habitats have involved many people, and several have been set up on the ocean floor at one time. The emphasis in one program may differ from that in another, but each has its place.

The way people view the ocean bottom differs and colors the world they see. Jacques-Yves Cousteau, who organized an underwater habitat thirty-six feet under the Red Sea in 1963, discusses the color and movement in the water, and tells how he loved to take out his small submarine at night to watch the sights, and how the men in his habitat had "a cosy sitting room, delectable meals, a luxurious bed and boon companions." [1] In contrast, Scott Carpenter, the astronaut turned aquanaut, spent a month at 205 feet off the coast of California in 1965. Working at a greater depth than Cousteau, he considered the deep ocean far less romantic than the moon. He writes:

I know that the moon riding high and bright in the clear night sky, is a beckoning symbol to men and always has been. I know that the ocean floor is a murky, cold unpleasant world that only dirty-handed, tough crazy divers can truly appreciate. Very few men will ever see the floor of the deep ocean themselves; it isn't *there*, like the moon.[2]

Yet this is the world in which some men are trying to live. Why? The most pragmatic reason is that the numerous oil installations on the continental shelf require people to repair and control them. Participants in Cousteau's second habitat worked on a seabed structure that duplicated many of the pipes, valves, and fixtures of a real oil wellhead. Navy divers like Carpenter have practiced underwater tasks, most of them directed to future military needs. Rescuing men from submarines, salvaging lost property, locating large and small articles on the bottom are some of the jobs Navy divers may be called upon to do. A variety of future installations, for military or civilian use, could be placed on the seabed. Tanks to store oil under water or ocean silos to hold commercial cargo may be used. When men begin to farm the sea, perhaps they will need to live near their fields of sea plants or herds of fish. Remote-controlled machines and submersibles with a crew inside at normal atmospheric pressure can do many ocean tasks, including much needed scientific exploration. But remote-controlled vehicles cannot exercise judgment, and mechanical arms on a submersible are clumsy. Some underwater jobs must be done by human beings. If casual visitors live in submerged hotels or visit aquatic museums, they will remain in the well-lighted, shallow waters off warm coral islands. The serious problems to be solved about ocean habitats come at the greater depths where the most difficult work must be done.

There are many problems involved in remaining beneath the water for prolonged periods. By 1958 scientists had already solved a number of them in theory. First, and most important, is the danger of the "bends." When a diver is down in the ocean, his body is under greater pressure than that of the atmosphere, and nitrogen from the air he breathes dissolves in his blood, eventually penetrating to the marrow of his bones. When he returns to normal atmosphere pressure, he must come up slowly enough to allow his body to decompress gradually, giving the nitrogen time to be washed naturally from his bloodstream through his breathing. If he comes back to normal pressure too rapidly, the nitrogen, not having time to leave his body through his lungs, will form bubbles like those in champagne. These bubbles of nitrogen, appearing in the bones and joints, cause what helmeted divers and caisson workers have traditionally called the bends. In many cases nitrogen bubbles cause crippling bone defects, the full effects of which may not appear until some time after the dive. If severe enough, the trouble may be fatal.

Years ago the British scientist John Scott Haldane experimented with the problems of decompressing men who had worked under water. He predicted the length of time required for safe decompression after dives of various depths. The United States Navy publishes decompression tables, which are used by many divers to help them calculate how long a time they must spend returning to the surface. In short, shallow dives, men can use a marked guide rope with stations at various depths, where they must wait patiently for the prescribed length of time before ascending farther. The greater the depth and the longer the time spent under pressure, the more time necessary for coming to the surface safely.

As diving became more common, scientists realized that if a diver stayed down for a long time, he eventually became saturated with nitrogen and no more could dissolve in his blood. Once that point was reached, it made no difference whether he remained on the sea floor for one day or ten: the same length of time would be required for safe decompression. This is called *saturation diving*.

Scientists also realized that divers did not have to wait for decompression to take place while they remained in the water. Decompression chambers, lowered into the sea, had already been used by navies in various parts of the world. On land, chambers were being built that reproduced the conditions found in the sea, and in them experiments could be run, first on animals, then on men, to test new theories. During tests on land, men could return gradually from great pressure, under controlled conditions and the surveillance of a physician. There was no reason why men could not be brought up from the ocean depths in a sealed container, to pass their long wait for decompression in a chamber such as those used in tests.

Another major difficulty for men staying under water comes from the mixture of gases they breathe. Oxygen makes up only about a fifth of the surface air. Therefore, under pressure, oxygen must be carefully controlled so that a man breathes the same amount he would receive on the surface. Too much of this vital gas affects the central nervous system, causing muscular twitching and convulsions. Too little is fatal. Nitrogen, an inert gas (about four-fifths of ordinary air), not only is dangerous to divers because it causes the bends, but is probably the cause of nitrogen narcosis, or the euphoria of the deep. Under certain conditions, divers become drunk with a feeling of joy; they forget what tasks they must perform; they lose track of time; and they have been known to throw away their scuba gear and drown.

This drunkenness seems to follow few rules and affects some people more than others. To eliminate some of the difficulties of nitrogen, scientists have experimented for many years with a mixture of oxygen and helium for divers. Helium, a light, inert gas, reduces the danger of narcosis, improves breathing under pressure, and has greatly increased the depth to which divers can go. It does, however, have its own difficulties. Fairly rare, helium costs a good deal and is not available in some parts of the world. It increases a diver's feeling of cold: breathing helium, he may be uncomfortably cold at 80 or 85 degrees. Helium also changes the human voice, making a man squawk with what is called the "Donald Duck effect." Although people at first found this an amusing annoyance, the need for divers to communicate with their surface ship and with each other is so important that efforts are now being made to develop telephonic devices that will unscramble the Donald Duck voice of the aquanauts.

Using this idea, Edwin Link, a man who had already made a fortune building ground trainers for aircraft pilots, designed one of the first diving chambers. Called the "submersible decompression chamber" (SDC), this cylinder could lower a man into the sea, serve as his house on the bottom, carry him to the surface under pressure, and then remain on the deck of a support ship while he was decompressed as slowly as necessary.

Link's first important experiment with his SDC was carried out in 1962 in the Mediterranean, off the French Riviera. The diver was a young Belgian, Robert Stenuit. An aluminum cylinder over eleven feet long, the SDC could be placed horizontally on the deck of a ship, but stood upright in the water. On the bottom, water did not rush into its open lower hatches because the internal pressure was equal to that outside, and kept the water out. After Link had tested the SDC at several depths, Stenuit went down, intending to remain at 200 feet for forty-eight hours. He was to swim around outside the chamber part of the time, and then climb up high and dry into his little habitat.

Stenuit found life at 200 feet exciting but hardly comfortable. Since he was breathing helium with his oxygen, the water felt extremely cold. In his elevator-like house, he could put on dry sweaters and use the heater that was on a top shelf. His upper body was then too warm, and his legs and feet were chilled. The atmosphere was dripping wet, and under pressure his air mattress acted strangely. When food was sent down to him, the container leaked: he ate cold sea-water soup, soaked French bread, and salty cake.

None of the problems was serious until a mistral blew up on the surface. This sudden Mediterranean storm produced turbulent seas and swamped an auxiliary boat bringing extra bottles of helium to the support ship. Since Stenuit was breathing an oxygen-helium mixture, he could not switch to oxygen-nitrogen until decompression had been completed. Although he was not told about the loss of the helium bottles, he obeyed orders to end the dive immediately. The chamber, with the diver sealed in, was raised to the surface, where the wind and waves tossed it about until it was securely fastened down on the ship's deck. Over the next two days, Stenuit was decompressed under the supervision of a doctor. The extra helium arrived in time from an emergency supply, and he suffered no ill effects from the 200-foot dive.

The following year, Edwin Link worked closely with the United States Navy in preparing a more comfortable living arrangement for men under the sea. This new habitat was called SPID, the "submerged portable inflatable dwelling," which, after much experimentation, took the form of a tent-like structure. It had a rigid frame and could be blown up under pressure. The cylindrical SDC was still used to carry divers to the surface under pressure and as a safety back-up device under water. On the deck of the support ship was a comfortable decompression chamber that could be mated with the SDC, allowing divers to climb out of the cramped space of the cylinder and to spend their long wait for decompression on full-length bunks in the deck chamber.

For this second Man-in-Sea program, Stenuit was joined by Jon Lindbergh, the son of Charles A. Lindbergh, the first to make a non-stop solo flight across the Atlantic. After considerable delay, Link put down his habitat near Great Stirrup Cay in the Bahamas in 1964. He kept his divers in their underwater home for forty-nine hours at 432 feet. It had been hoped that Stenuit and Lindbergh would be freed completely from dependence on the surface support ship, but because of mechanical difficulties in the SPID, tools and replacement parts had to be sent down to them. Compared with the original chamber used in 1962, the SPID was very home-like, although it was still damp enough to cause skin softening and rashes. Both men wanted the temperature at more than 82 degrees because of their helium-induced heat loss. They felt that they needed heated diving suits even in the warm waters of the Caribbean. Working outside their inflated house for several hours each day, the men found that they tired quickly. Inside, they could not communicate with each other because of their Donald Duck voices. Their sleep was disturbed by the cold and by a regular thumping against the wall of their dwelling. A large, stubborn grouper, chasing small fish attracted by the light, ran into the SPID every time he attacked his prey.

After 92 hours of decompression in the deck chamber, Stenuit and Lindbergh emerged into the normal atmosphere and found themselves being greeted in Miami by enthusiastic reporters. One advantage of a deck chamber is that the ship can return to port while the aquanauts wait for decompression to be completed.

The second Link experiment proved that men can work and live successfully at 400 feet. But many improvements are still needed. The men's comfort and safety were increased by the use of three different pressurized units: the deck chamber, the SDC, and the SPID. When the SPID had mechanical difficulties, the SDC provided a safe retreat. The problems of humidity and cold were not solved, and the divers' inability to understand each other's words was a great handicap. Undoubtedly, however, men will be able in the future to work at 400 feet without frequently coming to the surface.

At one time, Jacques-Yves Cousteau considered joining Link in his experiments, but the two men found that their aims were fundamentally different. Consequently, and almost simultaneously, both men were establishing habitats. Cousteau, at first, was more interested in setting up practical, comfortable living arrangements in the sea than in proving that divers could work at great depths. His first project, in 1962, called Conshelf I, put two men, for a week, in a steel cylinder at 32 feet off Marseille, France. With this experience behind him, Cousteau undertook, in 1963, a major experiment called Conshelf II.

The heart of Conshelf II, located 36 feet down on a coral ledge in the Red Sea, was the Starfish House, a submerged dwelling with four outer rooms and a large central salon. In one of the rooms, containing the kitchen and laboratory, the chef prepared delicious dinners with wine and cigars. When the lights were on, countless fish moved back and forth in fascinating patterns outside the picture windows. In the salon, men off duty played chess and smoked cigarettes. A parrot named Claude, which had been carried down in a pressure cooker, lived with the men to act as a safety check on their breathing mixture. If the air was unsafe, Claude would faint long before the men were bothered by it.

What difference was there between the Starfish House and Link's SDC in which a man lived cramped into a decompression cylinder for twenty-four hours? Instead of facing the dark and cold of 200 feet, the men in Conshelf were at only 36 feet, under twice the normal surface pressure, breathing compressed air. Two of Cousteau's men experienced the difficulties that faced Link's divers. While five men lived in Starfish House at 36 feet, two divers went farther down to spend a week at 90 feet in a cylindrical chamber, the Deep Cabin. Breathing oxygen and helium, they were to swim down to do useful work at 165 feet and to make short dives below 300 feet. They actually reached 363 feet, although they could not remain there. At this greater depth, they were bothered by squawking voices, humidity, ear pains, and sleeplessness. Oddly enough, however, their living quarters were too hot rather than too cold. The Red Sea was at 86 degrees in that area. Then the men came back to comfortable quarters in Starfish House and could thus return to normal pressure in stages.

Cousteau housed his small submersible, the *Diving Saucer,* in an underwater garage near the rest of the Conshelf II establishment. Inside the garage, the flat saucer-like machine could be hauled out of the water. Because it was at normal atmospheric pressure inside, Cousteau and a second man could take it down to 900 feet to enjoy the delicate pastel coral and the wriggling sea life that he photographs so well.

Conshelf II proved that life under the shallow sea could be civilized and comfortable. Cousteau chose his men for their special skills rather than for their age, stamina, and diving experience. At that time he believed that practiced divers were not necessary for testing life in a habitat. If human beings eventually have to live under water to service oil wells or shepherd fish, it is to be hoped that their life will be as comfortable as that of Starfish House. Just below the surface is a satisfactory place to live, for wind, waves, and weather do not reach down 36 feet. However, Cousteau demonstrated that comfort is also associated with support from a surface ship, with sunlight and warm water, and with pressure not too different from that on land. Men in Starfish House had sunlamps and French cuisine; in Cousteau's Deep Cabin, conditions were quite different.

Cousteau's next project, Conshelf III, concentrated more on working at depth than on the pleasure of living under the sea. In 1965, off Cap Ferrat in the Mediterranean, six men descended to 330 feet to live in a 140-ton spherical house and to work on machinery like that used in underwater wellheads. The *Diving Saucer,* with its cabin at normal pressure, went down to facilitate certain things, but there was no longer any question of casual visitors from the surface.

The house itself was less dependent on support from above than the previous habitat had been. The lines coming into it carried only telephones, television, and other electrical services. The breathing mixture was made from gases stored in tanks on the bottom. The use of helium produced the usual problems of cold and unintelligible voices. That gas was also to blame for other sources of bother to the men. Because the atomic weight of helium is very low, the gas seeps into very small spaces. The wristwatches of the crew, for instance, exploded when they surfaced because helium, under pressure, had been trapped inside them.

Because pressure ruins the insulating properties of most wet suits by crushing their little air spaces, Cousteau's men tried out a vest made with tiny plastic spheres embedded in rubber, each of which enclosed a minute amount of air. The vest improved working conditions, but the men, after working in the water, still were glad to return to the 90-degree temperature normally maintained in the habitat.

Unlike the workers in Conshelf II, the six men in Conshelf III were carefully picked from fifty candidates, and they proved to have exceptional stamina. The leader, at thirty-seven, was the oldest. The underwater wellhead used in the experiment was a standard type for oil wells, a structure called a "Christmas tree." Compressed air was used inside to simulate conditions found in a wellhead under pressure. Watched by oil engineers via television, the divers performed tasks that future underwater workers at real drilling sites

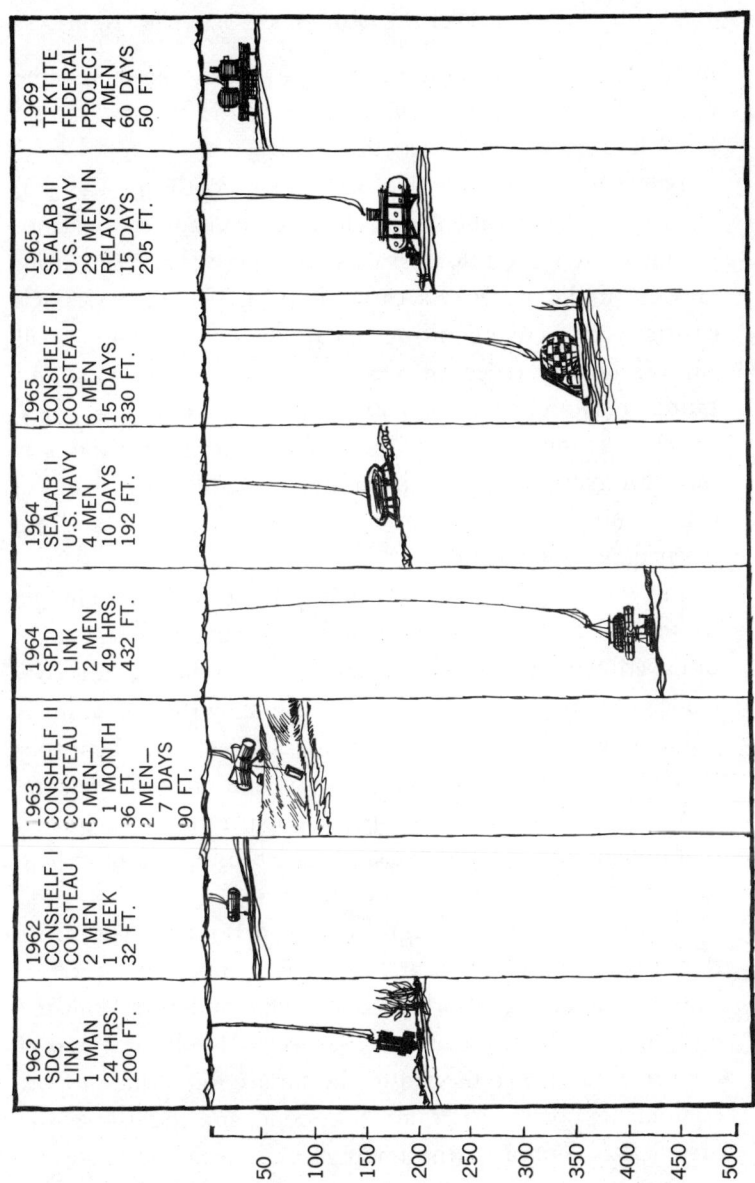

might have to carry out. After three weeks, the men came up, locked in the house, which rose to the surface as soon as its lead ballast was released.

Before men tried to live at various depths in the sea, scientists had experimented with pressure chambers on land, producing test conditions exactly like those that result from pressure under water. As early as 1923, United States scientists were studying nitrogen narcosis by observing small animals in an oxygen-nitrogen atmosphere. Mice, rats, and guinea pigs are often used in such work, as well as goats, which have much the same blood and breathing apparatus as man. Monkeys are not used. Though more like men than the other animals, they contract pneumonia too easily and are often difficult to handle.

Just before Edwin Link's first project, Captain George Bond, a U. S. Navy physician, had been carrying out a series of experiments reproducing deep-diving conditions in a controlled decompression chamber. Beginning with small animals going down to the equivalent of 200 feet, he then subjected Navy divers to the same conditions of pressure, temperature, and humidity. In these experiments he proved that human beings could live at great depths in an atmosphere of oxygen and helium without harm.

By 1964 there was enough interest in Bond's experiments to persuade the Navy to attempt a real underwater project. This was Sealab I, in which one officer and three enlisted men stayed for ten days at 192 feet. The habitat was a long steel capsule connected to the surface for all life-support equipment: power, water, air, telephone, and television. The first Sealab confirmed the findings of Link and Cousteau—that men could live at considerable depth, but that the art of doing so had not been perfected.

The second Navy project under Bond's direction, Sealab II, was one of the best-organized and most ambitious of the underseas habitats. In 1965, in cooperation with the University of California's Scripps Institution of Oceanography, the Navy put down, off the California coast, a steel cylinder, 12 by 57 feet, that looked much like a railroad tank car. It contained a scientific laboratory, a galley and dining areas, bunks, and other necessary space for ten men. There was also a chamber for carrying men under pressure to the surface, although the support ship had the usual decompression chamber. The depth chosen was 205 feet, where the oxygen-helium breathing mixture was necessary. Twenty-nine Navy and civilian divers were used in relays. Each team of men stayed down fifteen days out of the total period of six weeks. The one diver who stayed in the habitat for a full month was the astronaut Scott Carpenter.

The men of Sealab II lived a fairly Spartan life, eating freeze-dried food, supplemented by fresh meat and cake sent down from the surface. They slept in rows of double bunks —each man in a lower berth having a porthole to use for fish-watching. Although they could have watched commercial television, they preferred to watch real sea lions dive to catch fish attracted by their lights. They had a trained porpoise named Tuffy, who carried mail and other small items down to the habitat. Tuffy had been trained, on hearing a particular distress signal, to carry an airhose to a diver in trouble. In the habitat, the men had no cigarettes because matches would not light in the helium atmosphere, and cooking was complicated because water would not boil until it reached 300 degrees. Eggs were dangerous because they gave off a gas difficult to filter from the air, and fried food produced a stubborn greasy odor. Every day a doctor on board tested each diver's heart, blood pressure, sight, hearing, strength, and coordination. Bond observed the men on closed-circuit television.

Forty-seven different experiments were scheduled for the six-week period that Sealab II lasted. While the Navy was primarily interested in salvage and engineering work, there were also programs in biology and oceanography. Divers studied caged fish, the reaction of plankton to light, and the behavior of crabs and starfish. They made observation trips to Scripps Canyon, a deep area in that part of the ocean. Near the end of the mission, the men used a foam mixture to raise test drums to the surface, and raised an airplane hull from the bottom. They patched parts of submarine hulls, practicing the kind of work that would be necessary if the Navy tried to bring up a disabled submarine. Only one injury occurred in the water during the program. Carpenter was stung by a scorpion fish and kept out of action for twenty-four hours.

Under strict Navy supervision, with Bond in charge of the men's health, Sealab II was a well-conducted test in underwater living. The complex scientific results were carefully analyzed and correlated before they were made available for future use.

A third Sealab effort was planned for late 1968. Scott Carpenter was appointed second-in-command on the surface ship. He could not dive, for dead spots had developed in both of his thigh bones. These scarred and calcified areas, which show up as white shadows on X-ray photographs, were produced by a minor case of the bends. In spite of the great care used in decompression of the aquanauts in Sealab II, Carpenter at one point had severe pains in his legs. When the pains stopped, they were forgotten—until the ominous white shadows appeared on his X-rays. If the dead spots had been in the joints, he would have been crippled. As it is, he can engage in most normal activities, fly an airplane, drive a sports car, and play tennis, but he has been warned against further saturation diving.

With the development of plans for Sealab III, underwater living ceased to be a daring experiment and became a costly organized effort, run by the government for national purposes. Various organizations, including the Bureau of Commercial Fisheries, the Philadelphia General Hospital, the Navy's Mine Defense Laboratory, and the Undersea Warfare Center planned projects in connection with Sealab III. Divers were expected to construct a small building under water and to evolve techniques for finding lost objects on the sea floor. The cost of the program was originally estimated at 10 million dollars. Plans for Sealab III give some idea of the complex scientific and technological organization needed for a major undersea program, and suggest what military and civilian groups expect from the sea in the future.

New technical advances were ready for Sealab III. The problem of the terrible cold, especially at the planned depth of 600 feet, was to be overcome with heated wet suits, of which several kinds were to be tested. The men were to spend twelve hours in a chamber on the support ship's deck, gradually exposed to pressure until they were ready to descend in a personnel transfer capsule. Then they would spend a prescribed time living in the habitat and working out in the water. When they returned to the surface, it would take six or seven days for them to return to sea-level pressure safely. The main habitat was still the tank car, now enlarged by several extra rooms, one of which contained hot showers to be used by divers to warm up after their period of work outside. Although no one predicted that the teams of aquanauts would live in luxury in the bitter dark and cold of 600 feet, many new advances had been perfected to overcome the difficulties they would encounter.

This has not all worked out as planned. Sealab III has had more than its share of mishaps and more than the usual problems that arise in testing and organizing new equipment. Wiring defects in the electric motors aboard the support ship, *Elk River*, though not serious, caused delay. Leaks were found in the piping system for the helium breathing mixture. Both the deck decompression chambers and one of the personnel transfer capsules sprang leaks.

Since helium penetrates many conventional seals, such defects are not surprising. Each leak had to be corrected, however, although it meant time was lost. During training exercises, one of the transfer capsules flooded with water. In a complex project, some of these things are expected, but such delays add to the cost.

In mid-February 1969, Sealab III's main habitat structure was lowered to the sea floor, and the first aquanauts entered pressure chambers to prepare for diving. Since gas appeared to be leaking from two places in the habitat, four divers were sent down ahead of the main team to make repairs. Then disaster struck. Berry Cannon, a civilian diver from Panama City, Florida, had some kind of seizure. (His actions were clearly visible on a television monitor on the surface.) Although his companions rushed him into a transfer capsule and massaged his heart during the half-hour trip to the surface, he was dead when they arrived.

In the first reports, Navy doctors said death was due to

cardiac arrest, but were unable to give any clear reason for this. After the confusion caused by the event was over, one breathing apparatus on deck was found to have a defective unit for removing carbon dioxide. The question of whose responsibility it was to inspect or repair this equipment was raised. Later reports confirmed that Cannon had been asphyxiated by carbon dioxide, but added that extreme cold, caused by breathing helium, was a major factor in his death. It is possible that had Cannon not been so cold, he would, as an experienced diver, have recognized the defect and sought help in time.

Chiefly because of Cannon's death, further experiments with Sealab III have been postponed until some final decision about its future can be reached.

Despite the difficulties encountered by Sealab III, it seems that living under the sea at shallow depths is becoming a widely accepted, almost routine matter. Project Tektite, a habitat for four scientific aquanauts, was put down in early 1969 off St. John, Virgin Islands. It was at a depth of only 50 feet, in warm water, where conditions were relatively safe and techniques perfected earlier could be used. The Interior Department, NASA, the Navy, the Coast Guard, and the University of Pennsylvania were all involved. A private contractor, General Electric, built the basic habitat structure. After living in the habitat for 60 days, the diver-scientists returned to the surface in excellent condition, reporting that their morale had remained high during the long period in confined quarters. They felt that many more months could be spent on scientific work in that particular seabed area and that underwater living could benefit research in other regions of the ocean.

A second Tektite project began in the spring of 1970 at

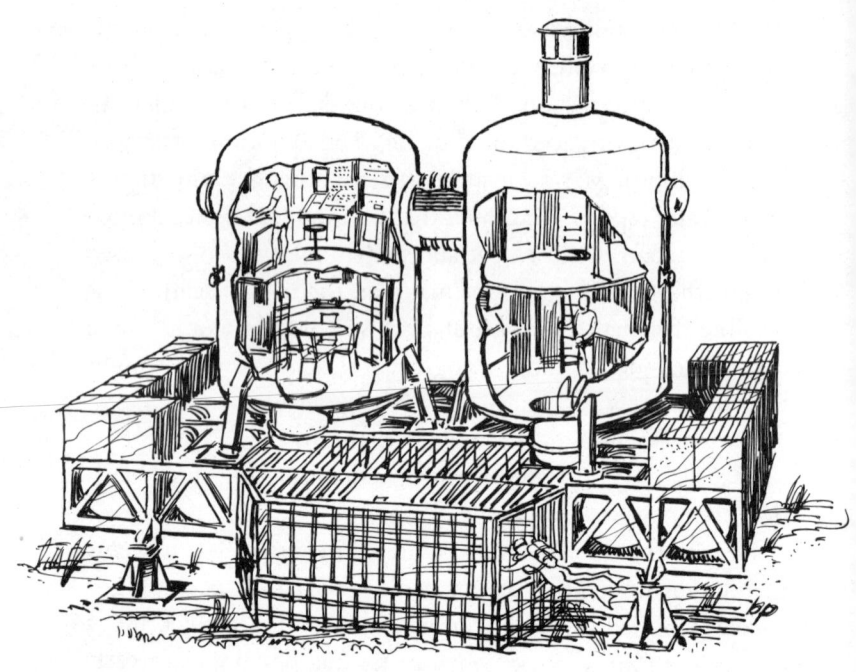

TEKTITE I HABITAT

the same location in the Virgin Islands. It will use teams of divers in relays to study marine life, seabed ecology, and the behavior of men living in confinement under stress. One team of divers will include five women scientists, and foreign divers have been invited to participate.

On land, tests have continued to see how far, theoretically, men can go down into the ocean's deep pressure. At Duke University divers have successfully reached a simulated depth of 1,000 feet and suffered no ill effects. Two of the men imitating deep dives complained of minor aches in their joints; but, except for this reaction, there seems to be no practical reason why men will not be able eventually to live at such depths.

There is always a gap between what man can endure in theory, or in a single limited test, and what he can endure over a period of time. There is also a gap between the first primitive machine, like Edwin Link's submersible decompression chamber, and a safe, working, manageable unit that will support ten or fifteen men. Small problems like designing a light bulb that will not implode too easily under pressure take a great deal of time and money. Cousteau figured that it cost him 250 dollars a day to keep Conshelf III supplied with television tubes because they broke so easily. Gradually the hardware must be developed at a low enough cost to enable private companies in the oil industry, or in future ocean businesses, to place divers where they need them. Techniques must be improved until they become safe and routine.

Although the death of Berry Cannon has cast a shadow over all divers, men continue to be willing to dive. The Navy had plans for Sealab III that cost millions of dollars, but Cannon's death was a major setback, causing investigation and delay. Nevertheless, experiments in saturation diving continue. The oil industry, the Navy, and scientific-research organizations are looking forward to a time when men can work successfully under water for long periods. If complex installations—from cargo silos to atomic power plants—are built on the ocean floor, the experience of living under the sea now being gained will become of vital importance. Link, Cousteau, Bond, and Carpenter are pioneers in a world that may become of more importance to us than we now think possible.

CHAPTER 8

The Uses of the High Seas

Four and a half miles off the southeast coast of Florida, there is a shallow coral area of the seabed known as Triumph Reef. Thinking that such an unoccupied spot in the sunshine should not go to waste, Louis Ray decided to build an island there, using hydraulic dredges to fill in the land. He intended to build up the area, occupy it, defend it to the best of his ability, and therefore claim and own it. He was going to call his new island the Grand Capri Republic. He himself asked the question: Would he and his four investment partners then become a foreign nation?

On the same area of the seabed, another development company had even more ambitious plans to build an island, called Atlantis, Isle of Gold, complete with a radio-television station, a post office, commercial buildings, foreign offices, a government palace for a new congress, an international bank, and a mint. If this too were a foreign country, then gambling casinos might be built, which would be highly profitable so near the United States shore.

Because these proposed new islands were outside the three-mile territorial waters whose seabed is owned by Florida, the United States federal government took the responsibility in 1965 of bringing a court action against the two companies. A decision handed down in 1969 states that, although the government has not taken formal title to these reefs, they are United States property, and the islands cannot be developed without permission from United States Army Engineers, who control all structures in United States waters that might impede navigation.

This is an indirect method of enforcing claims of the federal government to the seabed of the continental shelf, but it was apparently the only immediately available basis for preventing this unexpected use of the ocean. In the opinion of the judge, if these reefs were open to private construction totally outside the control of the United States government, not only unpoliced gambling casinos but also alien missile bases might be set up four and a half miles off shore.

The State of Florida has since passed a bill protecting these particular coral reefs by labeling them "natural resources." Federal law is not completely adequate to deal with possible future structures on the seabed because, although the court can stop such unwanted installations, there is no procedure for granting permission to build desirable private facilities in the future. Under the Outer Continental Shelf Lands Act of 1953, the Secretary of the Interior may lease portions of the seabed for utilization of oil and minerals, but for no other purpose. Although alien missile sites near the coast would be undesirable, underwater museums, skin-diving areas, or resort hotels might be assets, and some day may be built on coral reefs. If so, additional laws will have to be written.

Not all unwanted structures on the seabed can be controlled by judges using existing laws. In December of 1964, fifteen men of the Royal Dutch Navy and Police—ten of whom were dropped by helicopter—attacked and occupied a platform in the North Sea six miles off the coast of the Netherlands, just outside the territorial limit. Built like an oil rig resting on the bottom, this platform housed the broadcasting facilities of a pirate radio-television station, TV Nordzee. This station, in its three months of operation, had become tremendously popular with the Dutch people, had collected an estimated one million dollars in revenue from advertising, and had caused such a political dispute that a prime minister was forced to resign.

Previously, Holland had had two non-commercial television channels, both owned by the government, and directed by five separate organizations that represented the political and religious divisions of the country. Two were Protestant, one Roman Catholic, one Socialist, and one neutral. The government collects a ten-dollar tax on each of the nation's two million TV sets and apportions it according to the size and influence of each of the five organizations. In 1963, Dutch businessmen complained of the effect of advertising that reached the country from German stations. They asked Parliament to add commercials to the state-owned channels. When the request was rejected, a number of important Dutch businessmen started TV Nordzee.

TV Nordzee was not a full-time television station. It began broadcasts in the early evening when the other channels were not operating, was silent between eight and ten P.M.—the prime hours when the five regular organizations had their best programs—and returned to the air waves between ten and eleven, when movies were shown. In the few one-minute spots available for advertising, commercials were presented for Coca-Cola, Max Factor, Pan American, etc. This form of television became so popular that the Prime Minister of the Netherlands took up the cause of advertising on the state-owned channels, but he was defeated on the issue. As a result, he resigned.

The Dutch Parliament passed a special law, allowing the government to control such offshore structures on the continental shelf, even if, technically, they were on the high seas. Immediately the navy moved in, the helicopter lowered men onto the pirate TV platform, and the operators gave up, as expected, without resistance.

Unlike some of the pirate radio stations that have sprung up near the English and Swedish coasts, TV Nordzee was backed by reputable Dutch interests, although listed under foreign ownership. In reacting to it, the Dutch government passed a law that raises serious questions about the use of the high seas. If the Dutch own the seabed of the shelf off their coast, do they have control over any structure built on it? Would they also control a floating structure that was not embedded in the soil? The 1958 Convention on the Continental Shelf does not offer clear answers to such questions.

The structures off the Florida coast and in the North Sea off Holland were both just beyond the three-mile territorial limit—obviously put there because of the proximity of the coastal country. In each instance, one particular nation had a special concern in the area because it already had jurisdiction over the seabed to the edge of the continental shelf. But what of areas of the high seas neither on the shelf nor adjacent to a coastal country?

In the deep ocean, beyond the continental slope, great ranges of mountains and scatterings of volcanic peaks rise many thousands of feet from the sea floor and come to various depths beneath the surface of the water. Scientists, mapping and charting the ocean, have discovered a number of these seamounts that come near the surface—off the New England coast, in the West Indies, in the Sea of Japan. One extinct volcano, the Cobb Seamount, discovered by a member of the United States Fish and Wildlife Service in 1950, lies only 270 nautical miles off the coast of the State of Washington. It is definitely beyond the continental shelf but near enough to be reached easily from shore. Its environment is typical of the deep ocean, and yet its terraced summit comes to within 122 feet of the surface, giving it enough sunlight to support a thriving world of fish and marine plants. The area is untouched by previous human influences, a condition not found in coastal waters that have been fished over, polluted, and used as dumps.

A group of government, university, and scientific organizations, mostly from Hawaii and the State of Washington, in 1968 developed a plan called Project Sea Use, to put a habitat on the Cobb Seamount and make comprehensive studies of its environment. The blueprint calls for a habitat manned by four specially qualified scientist-engineers to remain on the mountain terrace for forty days. A mast is to be erected on the summit and left in place to collect data after the program is completed. This temporary mast may subsequently be replaced by a permanent tower. One or two underwater vehicles are expected to aid the divers, and the habitat will be supported by several surface ships.

Only the preliminary stages of this project have been carried out. In October 1968 a United States Coast Guard and Geodetic Survey ship, the *Oceanographer,* made an intensive study of the underwater mountain, surveying the terrain with sophisticated electronic equipment while teams of divers explored the upper slopes, observing red snapper, rockfish, and a variety of plants. The next step is scheduled to be a study of the area by a new submersible, the *Deepstar-2000,* which has been undergoing sea trials off the coast of California. Further events in the planned program will probably depend on what funds are available for Project Sea Use and on the lessons learned from other habitats.

The planners of Project Sea Use, writing about the Cobb Seamount, emphasize the new knowledge that can be gained from it—information about the atmosphere, ocean chemistry, the biological environment, currents, wave motion, water properties, and dispersed pollutants. They are also careful to consider the legal precedent that may be set if the Cobb Seamount is occupied, even temporarily. Outside both the territorial limit of the United States and the continental shelf, the underwater peak is not covered by international laws concerning those areas.

The planners of Project Sea Use write that the seabed of

the deep ocean can be considered *res nullius,* belonging to no one. Just as nations in the past have acquired sovereignty over land areas by occupying and settling them, the United States may well establish jurisdiction over the Cobb Seamount by occupying it and will certainly prevent anyone else from doing so. Possibly, in the future, the seabed will be placed under some shared authority, as land was in Antarctica. Another alternative is for the United Nations to control and exploit it for the benefit of all nations. In that case, write the men who intend to explore the Cobb Seamount, the United States may still be allowed jurisdiction over this particular area if temporary occupation has taken place.

With the states claiming the seabed under territorial waters and the federal government owning the continental shelf, it is uncertain what governmental agency would have the power to encourage or restrain the organizations about to occupy a seamount. As new technological developments such as submersibles and habitats make exploration of seamounts possible, the Law of the Sea will be affected by every event on the ocean floor.

On the continental shelf and crisscrossing the high seas, the United States has one of its first lines of defense, the nuclear submarines that carry ballistic missiles, and the antisubmarine devices that warn of any threat from foreign underseas vessels bearing the same weapons. The first nuclear submarine, the *Nautilus,* was launched in 1954 after Admiral H. G. Rickover had persuaded the nation to put 250 million dollars into a five-year program to create a nuclear submarine force. Unlike diesel submarines, which must surface regularly to reach the oxygen needed to burn their fuel, modern nuclear ships can remain under water for extended periods, returning to port only to meet the needs of the crew. On the surface a submarine can easily be located by an enemy, but under water it is harder to detect. High-frequency scanning methods, using light or radar, cannot penetrate water effectively. This leaves submarines free from detection except by less efficient sound devices, such as sonar. Nuclear submarines, constantly submerged, cannot quickly be located and rendered harmless.

Soon after the success of the first nuclear submarine, the United States Navy developed the solid-fuel Polaris missile and overcame the problems involved in firing such a weapon from a vessel under water. These Polaris missiles, some of which can carry a nuclear warhead for 2,500 miles, have been followed by an even more effective weapon, the Poseidon, a 34-foot missile with a nuclear payload of almost three megatons.

In 1968 the United States had thirty-two nuclear attack submarines patrolling the oceans, and another twenty-four under construction. Many of these ships, based in Spain, Guam, and Scotland, have a crew of 140 men, remain at sea for sixty days, and carry sixteen missiles, each of which can be fired at a different target in an emergency. Although there are few official figures, it is known that the speed of these ships is over thirty knots, and they are believed to run at a level between 1,500 and 2,000 feet.

The United States Navy is planning a new fleet of supersubmarines that will travel at a lower speed and have a new propulsion system to eliminate noise, making them so quiet that their detection will be difficult. Costing 100 million dollars, compared with 83 million for previous nuclear ships, the new vessels will be 20 percent bigger. The first of the new design is expected to be in service by 1973.

The chief purpose of the Polaris-Poseidon submarine fleet is to keep at sea a striking force that would retaliate if any nation were to attack the United States. If all United States land-based weapons were put out of action during such an attack, it would still be impossible for the attackers to find and destroy every nuclear submarine capable of striking back.

While the United States has been developing this vast underwater deterrent system, the Soviet Union has been working on a somewhat similar force. Few figures are available about the size of the Soviet submarine fleet, but the U.S. Defense Department announced in 1970 that a Soviet missile-carrying submarine was patrolling in the Atlantic.

Researchers in the United States estimate that the Soviet Union has not attempted to track all the Polaris submarines cruising the ocean. Until 1968 the United States used a tracking system to follow the movements of all Soviet submarines from the time they left port until they returned again. As the Soviet vessels increase in number, become faster and more enduring, the cost of tracking them increases proportionately, and the United States Congress was unsure, in 1969, whether such efforts in the future would be worth the money. If Soviet nuclear submarines can fire from 2,500 miles away, a new expensive system of warning devices may become necessary unless the United States gives up trying to know where every Soviet vessel is.

A major part of the United States' existing defense against foreign submarines is a warning system called SOSUS, a series of listening posts strung along the ocean floor in strategic locations. These are chiefly passive sonar with open microphones that send out no sound, but merely pick up noises within their range. The sound issuing from any submarine is so distinctive that a sonar expert can usually distinguish between two ships he has heard before, and computers can be used to identify vessels by their particular hums and rattles. Whether this warning system will be effective against newer underwater vessels from other countries is still open to question.

When the question of military weapons in the ocean was discussed at the 1958 Geneva Conference, the United States opposed any ban on weapons, partly because its own submarine-detection devices might easily be considered instruments of warfare and the country would be poorly defended without them. Members of the Conference agreed to omit any reference to seabed military installations in the conventions, and to allow questions about nuclear weapons to be considered later in some conference about general nuclear armaments.

At a United Nations Disarmament Conference in Geneva in 1969, the Soviet Union and the United States disagreed at first about the type of weapon systems that should be outlawed in the sea. Representatives of the two nations finally agreed to a tentative proposal forbidding placement of nuclear weapons or other weapons of mass destruction "as well as structures, launching installations or any other facilities specifically designed for storing, testing, or using such weapons" [1] on the seabed or under it. This would include nuclear mines anchored or embedded in the ocean. It would not apply to submarines, nor to an antisubmarine warning system, a defensive system that the United States would be unwilling to forgo.

The fate of this draft treaty is uncertain because the

United Nations Assembly rejected it in December 1969, and was expected to return it to Geneva with comments. The Assembly also passed a resolution asking the United States and the Soviet Union to stop testing and deploying new nuclear weapons.

Part of the United States' ability to gather information rests on the use of electronic intelligence ships on the high seas, vessels like the *Pueblo,* which was captured by the North Koreans in January 1968. According to an official United States report sent to the United Nations, the *Pueblo,* with its captain, Lloyd S. Bucher, and a crew of eighty-three men, was twenty-five miles from the North Korean port of Wonsan and over fifteen miles from the nearest island. Since the United States has been willing to accept the North Korean territorial limit of twelve miles, the ship had orders to remain thirteen miles from shore. Presumably, it was on the high seas.

One interpretation of the incident suggests that the North Koreans may have been ready to seize any opportunity to harass and disgrace the United States, regardless of the *Pueblo's* position. Four North Korean vessels surrounded the American ship, fired on it, killing one man, and seized it with an armed boarding party. Bucher reported that he and his men unsuccessfully attempted to destroy secret information on board before they were forced to enter the North Korean port. When they were later freed, the crew of the *Pueblo* reported that Bucher was threatened with the death of his men if he did not sign a confession that the *Pueblo* had been a spy ship.

When the United States signed a confession of espionage, the *Pueblo's* men were freed after nearly a year's imprisonment. At the same time, the United States denied any guilt. When a Navy Board of Inquiry heard testimony on the case, the presiding officers were chiefly interested in finding out whether secret documents and equipment had fallen into enemy hands. There was little discussion about the exact position of the ship when it was seized.

Arthur J. Goldberg, then United States Representative to the United Nations, called the seizure of the *Pueblo* "an act of wanton lawlessness." [2] Since the *Pueblo* incident, the North Koreans have attacked a United States aircraft that was not infringing on their territory. Although these episodes involve questions about the uses of the high seas, their settlement has not clarified the law. The United States has negotiated on the question of espionage rather than on the limits of the territorial sea.

The seabed of the deep ocean—the thousands of miles of sea floor beyond the continental shelf—was not fully discussed at the 1958 Geneva Conference on the Law of the Sea. Provision was made for the seabed of the continental shelf, out to some boundary defined as a depth of 200 meters or any depth where man could exploit the natural resources. Rules were laid down for conduct on the surface of the high seas. But two questions remained: exactly where does the seabed of the deep ocean begin, and under whose jurisdiction is this vast area of the earth's surface?

In 1967, Arvid Prado, the Permanent Representative of Malta to the United Nations, introduced in the General Assembly a proposal for the peaceful uses of the seabed of the ocean. He recommended that such activities as the construction of military installations and the dumping of radioactive material be outlawed and that the entire seabed be under the direction of the United Nations, which would use any wealth derived from it for the benefit of the poor nations of the world.

How much wealth lies on the ocean floor, and how soon may it be possible to use that wealth? These are open questions. Certainly the area involved is tremendous, for water covers two-thirds of the earth's surface. Minerals exist there in quantity. Ambassador Prado states that there is enough aluminum to last the world for 20,000 years, enough manganese for 400,000 years. In 1969 a United States Navy research team found salt domes in the deep ocean 400 miles west of Senegal, Africa, indicating that oil also is probably there. But exploiting these resources depends on the state of technology, on the cost of ocean products compared with the cost of those found on land, and on miscellaneous political factors.

Some difficulty may arise if nations with little experience of the sea are continually told that the deep ocean is a new source of unlimited wealth, like the American gold that enriched Spain in the sixteenth century. If the reality turns out to be less enchanting than the dream, there may be angry disappointment. A realistic estimate of ocean wealth suggests that both advanced techniques and a large capital investment are necessary before anyone will really profit from the sea floor.

The possible military use of the seabed is as important as its oil and other minerals. Malta's proposal stressed that the ocean, like Antarctica and Outer Space, should be reserved for peaceful purposes. Small countries, not directly involved with the deep sea, may urge peaceful use of the area in order to discourage their more powerful neighbors from engaging in a nuclear arms race. Any holocaust resulting from underwater weapons would affect them as well as the belligerents. For the United States, discussion of Malta's proposal raises questions about its own military needs and the plans that the United States Navy may have for the deep sea.

The immediate official United States reaction to Malta's proposal was guarded approval, with a strong indication that the time was not ripe for such a broad design for the ocean. Present knowledge, officials said, was not great enough to enable the United States to take a definite stand on the question. Further study was necessary.

Unofficial reaction was sharper and more divergent. Men with a generally international outlook favored some arrangement like that suggested by Malta because it would prevent a struggle among nations to seize parts of the ocean. An international agency controlling the seabed might be able to discourage pollution and ensure a fair distribution of the sea's potential wealth. In 1966, before any question arose of putting the seabed under United Nations jurisdiction, President Johnson made the statement:

. . . under no circumstances, we believe, must we ever allow the prospects of rich harvest and mineral wealth to create a new form of colonial competition among the maritime nations. We must be careful to avoid a race to grab and to hold the lands under the high seas. We must ensure that the deep seas and the ocean bottoms are, and remain, the legacy of all human beings.[3]

Forces in favor of Malta's general policy were led by the Senator from Rhode Island, Claiborne Pell, who outlined a new, modified proposal to guide the United States to an international agreement.

Forces opposing any treaty to close the deep-sea floor to

national ownership and to military installations took more time to organize but proved to be strong in their opinions. Congressmen Thomas Pelly, from the State of Washington, said that any acceptance of Malta's proposal would be "the biggest giveaway in the history of America."[4] Congressman Bob Wilson of California said that "a world agreement on oceanography sounds good, but is a booby trap for the United States."[5] Men opposing the proposal argued that, if the deep-ocean floor were under an international agency, with benefits going to poor countries, there would be little incentive for nations or private investors to expend the enormous effort needed to exploit the area.

Ownership of the seabed is closely tied to one question: where the continental shelf ends and the deep ocean begins. Opponents of the Malta Proposal generally favor a continental shelf that extends as far as possible. Both for the United States oil companies now exploiting the seabed of the shelf and for the federal government, which benefits from oil leases, the wider the shelf allowed along the coast, the greater the profit.

It is possible to interpret the exploitability clause in the 1958 Convention on the Continental Shelf to mean that there is no limit to the shelf and that exploitation can continue out onto the floor of the deep ocean. What would happen if the seabed were divided up like a pie, each coastal country extending its jurisdiction until it met the territory of a country coming from the opposite shore? Theoretically, extending the rules set up in the 1958 Geneva Convention, every coastal nation, and *every island,* could claim territory out to a median line dividing the seabed in half. If tiny islands could control the division of the ocean, an odd distribution would result. France would control the Indian Ocean, Great Britain the South Atlantic. Portugal, with the Azores and Madeira, and Spain, with the Canaries, would hold much of the North Atlantic. In the Pacific, the island of Midway would put the Soviet Union and Japan at a great disadvantage.

No one really expects this distorted division of the sea floor to take place. But many people, opposed to international control of the ocean, want a regime in which nations would be free to take riches from the seabed and to claim the areas they develop. At the same time, military forces in the United States are very hesitant about any international agreement that might inhibit their plans for the defense of the nation.

In the United Nations, the proposal by Malta led to the establishment of a temporary committee which was to study United Nations activities regarding the seabed and to indicate practical means of promoting international cooperation in this area. During a series of meetings, the committee reached no real agreement on a regime for the seabed. However, it accepted a proposal made by the United States that the 1970's be declared an International Decade of Ocean Exploration. This period of special emphasis is to stimulate investigation of the seabed and encourage cooperation in developing its resources. The committee also recommended that the United Nations establish a permanent seabed committee. This was done.

The United Nations again considered the seabed question in December 1969. It passed a resolution saying that, until an international regime has been established for the seabed, nations are to refrain from exploiting resources beyond the limits of national jurisdiction. Some experts fear that this may encourage further claims to national jurisdiction. The United Nations also authorized a study of whether its members wished a new conference to revise all subjects in the 1958 Geneva conventions.

While the United Nations was considering the question of the deep ocean floor, an important group in the United States spent two years studying and making recommendations about the entire scope of the nation's effort in the sea. This was the Commission on Marine Science, Engineering and Resources, created by an act of Congress in 1966. In 1969 it issued its first report, entitled *Our Nation and the Sea* (often called the Stratton report). Unlike many official documents, which are filed away and read only by experts, this study is likely to have a strong and long-range effect on United States policy concerning the ocean. It is both an excellent résumé of present conditions in the sea and a forecast of what may be expected in years to come.

The commission recommends that the United States initiate a new international agreement about the width of the continental shelf. It proposes that the outer limit of the shelf be fixed at a 200-meter depth or 50 miles from the baseline, whichever gives the coastal nation a greater area of seabed. To be marked off on the best charts available for each country, this boundary would remain fixed regardless of future changes in the shore line or new surveys of the ocean. By adding the 50-mile definition the recommendation would satisfy countries that have no continental shelf, giving them a fair share of the sea floor. However, because it would allow only a narrow shelf for national jurisdiction, nations that have developed expectations of going beyond this point might oppose the recommendation. To satisfy them, the commission recommends an added "intermediate zone," an area going out to the 2,500-meter depth or a 100-mile limit, in which the coastal nation would have special, but not exclusive, rights.

To control the seabed beyond the redefined continental shelf, the commission recommends that an international authority be established to register national claims, recording the particular area and the mineral resource to be exploited. The size of the area to be covered by a claim and the length of time the claim would remain valid are details that the authority would determine. To cover the cost of the authority, every nation would pay a fee for its claim.

The commission's recommendations for an international agency are intended as a framework for some future agreement about the seabed rather than as a final solution on which the United States would insist. Since the Malta proposal, a number of detailed plans have been suggested for control of the deep ocean, but the commission's report contains the one most likely to be used as the basis for the United States position on the question.

The commission's study gives a number of other recommendations that may influence ocean activities in the United States. They range from a research and development program for desalinization to a continental-shelf nuclear power plant, which would test the feasibility of sending large-scale power to coastal cities from offshore generating stations. It recommends better management of the nation's "coastal zone," which includes both the saltwater coastline and the Great Lakes shoreline.

The commission would like to see a new governmental agency established to control all the activities connected with the sea. Such an agency would be called the National Oceanic and Atmospheric Agency (NOAA), the initials deliberately reminding us of the biblical Noah who was a noted seaman, conservationist, and planner. The agency might include: the Bureau of Commercial Fisheries, now in the Interior Department; the Environmental Science Services Administration (which includes the Weather Bureau); the Coast Guard, now in the Department of Transportation; the National Oceanographic Data Center, and the Sea Grant Program. The Sea Grant Program, established under the leadership of Senator Claiborne Pell of Rhode Island, provides funds for colleges and universities to carry out marine studies projects.

Regardless of what new arrangements are made within the United States for dealing with the ocean, certain international problems are of immediate importance. The question of how to define the outer limits of the continental shelf will be debated by the nations of the world, and it is hoped that some agreeable solution will be reached. A method of controlling the use of the deep-ocean floor, which is closely connected with any decision about the width of the shelf, will have to be found. Some type of United Nations agency or international registry seems likely. Many experts feel that, when nations find it to their economic advantage to settle a question like the use of the ocean floor, they are quick to find solutions to even the thorniest problems. As soon as the North Sea and the Persian Gulf became profitable areas, the surrounding countries managed to divide the seabed among themselves. Perhaps, as the value of the deep ocean increases and nations develop greater ability to exploit it, they will find it easier to agree on methods of administering its control.

Sealab II and III

58 The *Berkoni*, specially built mother ship for Sealab II, arrives off La Jolla, California, where it lowered the habitat to the bottom

59 Scott Carpenter, sitting on Sealab II, just before the habitat was sent underwater

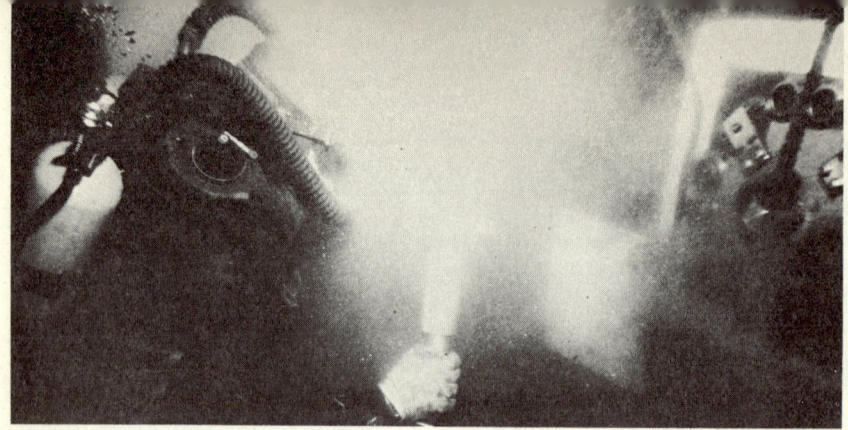

60 In a Navy salvage experiment, Aquanaut Billy Meeks patches a simulated submarine hull in the vicinity of Sealab II, 205 feet below the surface

61 Sealab II resurfaces after 45 days

62 Sealab III on a barge off San Clemente Island in late November 1968

63   On February 14, 1969, with its support ship in the background, Sealab III is towed into position for attachment of the umbilicals

64   On February 15th, Sealab III was lowered into the Pacific

65   On February 16th, divers checked out the rigging as Sealab III started to go below the surface

66   Berry Cannon, who died on February 17, after going underwater to repair a leak in Sealab III. This picture was taken in September 1968, as he prepared for a test dive in preparation for the Sealab III project

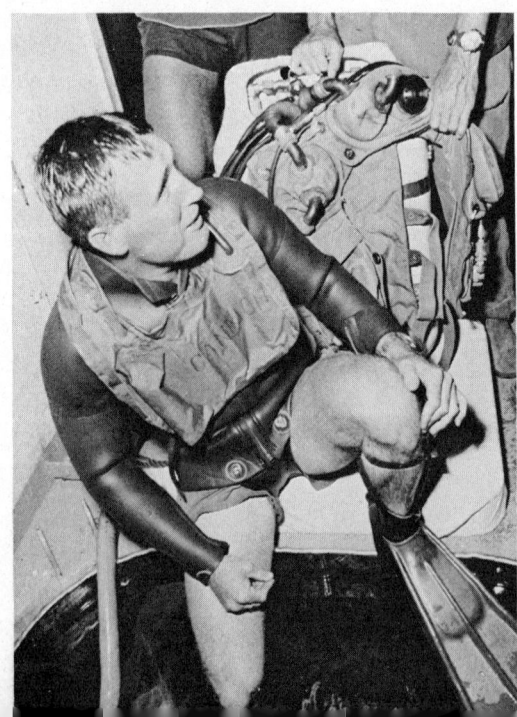

CHAPTER 9

The Future of the Sea

When we consider the future of the sea, we must decide not only what is possible but what will make best use of the ocean's great resources.

During the first half of the twentieth century, people felt that everything was possible to science and technology. This attitude resulted in a headlong rush into the Era of Affluence and the Space Age. Medical research lowered infant mortality and extended man's life span. Engineering provided labor-saving devices and new ways of enjoying leisure time. The automobile and the telephone decentralized cities and produced new suburbs and suburban industry, while ships and airplanes opened the world to middle-class tourists. Science explored the mysteries of the atom, of life-producing molecules, of the earth and the universe.

Only now is the affluent society beginning to realize the

cost of this tremendous upsurge in knowledge and standard of living. It has left behind large parts of the world's human population. Its wealth has been attained, and is maintained, at the expense of natural resources and other forms of life on this planet. Men are becoming aware that the uses of science must be governed by more than the immediate needs of one part of the earth's population. Science must benefit all people, especially future generations, if the latter are to escape living in a denuded, polluted world. Men are discovering that, while they live under the shadow of the atom, they also face the threat of gradual destruction of their environment.

By allowing DDT from the fields to poison nearby streams, and aluminum beer cans to lie along the roadside, men are constantly destroying the environment and fighting the natural forces that do the housekeeping of the earth. They are discovering that science is not as all-powerful as it seemed, since it cannot get rid of the DDT once it has been used.

The same pressures that created new problems on land will drive men to the sea for resources. The world's oil companies now spend as much money developing offshore oil resources as the United States spends on its space program. One scientist has calculated that it requires 4,000 barrels of oil (there are forty-two gallons in a barrel) to support adequate heating, transportation, automobiles, and electricity for each person in the United States for his lifetime. Because offshore oil produced in United States waters provides only 12 percent of our requirements, pressure for more offshore oil will increase as the land reserves are used up. This, in turn, will cause more oil spills that pollute beaches and kill shore birds. Conflict will increase over pollution of fishing waters, disruption of the sea floor, and obstructions to navigation.

Extensive surveys will be a first step in future ocean activities. Before intelligent use of the sea can benefit mankind—both present and future generations—more specific information about it is needed. The sciences of oceanography, of marine botany, biology, and geology are fairly new. Most current data have been obtained from surface ships. New abilities to work beneath the sea, coupled with modern instrumentation and computers to analyze data, will expand the range of collected knowledge.

Biologists want to observe marine life from seabed laboratories, and to use moving vehicles to provide observational facilities equal to those available to scientists on land. Geologists want to collect samples from a wide area of the ocean floor. Opportunities for such studies have, until now, been lacking; but now men can explore at first hand the fishing banks and coral reefs, the undersea mountains and deep canyons, the living organisms on the bottom, and the dynamic forces in the sea. They can track the rivers of the ocean, the warm Gulf Stream, the Kuroshio Current off the Japanese coast, and the Humboldt Current that surges across the Pacific.

The Gulf Stream Drift Mission of the submersible *Ben Franklin,* in 1969, was one much-needed ocean survey. Although millions of people yearly travel up and down the eastern coast of the United States, in cars, trains, ships, and airplanes, few know much about the immense river in the sea that influences the climate of both the eastern United States and Europe. Jacques Piccard has pointed out that, without the Gulf Stream, his native Switzerland would be submerged beneath a glacier, for the current's warm waters are Europe's heating system.

Another ocean survey is the Barbados Oceanographic and Meteorological Experiment, BOMEX. An area of the Atlantic Ocean, east of Barbados and north of the Equator, named the Doldrums, is of unusual interest to scientists because hurricanes spawn there. BOMEX is a combined air-sea investigation to determine the thermal interaction between the air and the water in this ocean area. The operation is being coordinated by Environmental Science Services Administration (ESSA) with the assistance of other governmental agencies, and combines the use of surface ships, aircraft, and satellites. Typical of the sophisticated research equipment used in BOMEX is a vessel called *Flip,* the Floating Instrument Platform. Built like a long, narrow ship, *Flip* can be towed to its destination. Once there, it is upended with its stern deep in the water and its bow up in the air. The stern becomes a very stable underwater platform for oceanographic instruments, far enough beneath the surface to escape wind and waves, while scientists live and work in the bow. The only modern oceanographic vessels not used in BOMEX are submersibles.

Beyond the continental shelves and the continental slopes, vast ocean areas have never been explored. Some of the world's most spectacular natural phenomena exist under water, such as the mid-Atlantic ridge, a mountain range to excite the imagination of the most expert mountain climber. This chain of mountains extends from north of Iceland down to the Azores, then through the South Atlantic, where it curves east, around the Cape of Good Hope, into the Indian Ocean. The islands of Tristan da

Cunha, Ascension, St. Helena, and the Azores are the only surface evidence of these underwater mountain peaks.

Certain scientists believe that billions of years ago the continents of North and South America were joined to Europe and Africa, and that since that time the great land masses have been drifting gradually apart. The mid-Atlantic ridge is thought to mark the dividing line where the continents were once joined. It is an area of violent volcanic action. In 1961 the inhabitants of Tristan da Cunha were driven off their island by volcanic eruptions and forced to live elsewhere—many choosing Britain—until the upheaval subsided. Several new volcanic islands have burst from the sea south of Iceland in a spectacular exhibition of the forces working within the earth's crust.

Only in the past ten years have scientists discovered that the mid-Atlantic ridge is a continuous mountain chain. The primary means of exploring it has been the sound fathometer on surface ships. Long, wide basins in the ocean floor lie on each side of the mountains, which rise to an average height of 5,000 feet. Ridges, trenches, and canyons form a rugged seabed landscape about which there is little information. Future scientific surveys will include increased underwater exploration of the mountain range by submersibles and habitats. The task is so imposing, however, that it will match exploration of the Rocky Mountains or the Himalayan peaks.

Exploring the ocean floor beneath the Polar ice cap will be an equally difficult undertaking. In 1958 the nuclear submarine *Nautilus* traveled under the Arctic ice and discovered that the sea floor is covered with jagged mountain peaks. A major mountain ridge extends under the water from Canada to northeastern Siberia. The Arctic seabed is particularly interesting to scientists because the North Pole may once have been located in a different position, in which case ocean sediment might record an era of green plants where there is now ice. The edges of the Arctic Ocean are fed by great rivers on the slopes of northern Canada, Alaska, and Siberia, and may contain a wealth of minerals.

Past explorations of the deep ocean floor have been limited by their distance from home base and by the exigencies of the weather. Future surveys will have the benefit of air transport for supplies and of the means to work under water, where the weather cannot reach. Studies can then be made of much wider areas of the ocean, in greater detail than ever before.

For a hundred years, Chile and Peru have produced nitrogen fertilizers from the seabird droppings along their rocky shores; now they find it more profitable to harvest the anchovetas on which the birds feed. A great concentration of fish is attracted by the Humboldt Current, which travels across the Pacific and surges to the surface near the west coast of South America, bringing nutrients up from the ocean floor. Great ocean currents like the Humboldt help explain why certain areas are rich with life.

Much of the sea floor is covered with a layer of nutrients, including nitrates and phosphates, the result of dead marine life drifting down. Deep-ocean fishes, with their strange shapes and feelers, many with bioluminescent lights down their sides, are able to survive only because of food that descends from above, like lunch brought by an elevator to men in a mine. All food originates in plants dependent on sunlight.

Nutrients from the sea floor are brought to the surface

by upwelling currents that can be explained by well-known hydrodynamic laws involving the shape of the seabed and the characteristics of sea water. In addition to being dependent on a supply of food, marine life is extremely sensitive to variations in water temperature and salinity. Using this knowledge of hydrodynamics and marine biology, it may be possible to create artificial upwellings and to change currents that affect temperature and salinity, thereby increasing the food productivity of the ocean. Large forces of nature would be involved in these manipulations, and they must not be undertaken without careful study. Tampering with any natural balance may have unexpected effects.

A less massive approach to food production in the ocean is the development of fish farming and herding techniques, comparable to the centuries of animal food production on land. Still working as hunters and food-gatherers in the sea, people now use farming techniques chiefly for stationary shellfish such as clams and oysters. Fish farming may be an important future source of food, provided pollution can be overcome in rivers and bays. But no one yet knows how to conserve, protect, and improve production of roaming marine life—the small schooling fish, the open-ocean fish, and the mammals, such as dolphins and whales. The mammals, some of which have brains larger than man's, may help man survive, possibly being used as horses and dogs are used on land. Experiments in communication with dolphins are important, not only for basic scientific knowledge, but also for practical improvements in food-gathering.

Although some Africans and the Japanese have been farming fish in pools for centuries, experiments with artificial reefs and tidal pools have only begun in the United

States, where natural fish-breeding grounds, such as tidal wetlands, are being destroyed at a frightening rate. Bulldozers, dredges, and pile-drivers are threatening shore areas essential to marine life. However, many people in the United States are beginning to understand the importance of maintaining ecological balances to protect future resources. To save tidal wetlands, authorities must reconsider the priorities put on seashore areas, and experts must improve their productivity as breeding grounds for fish. Only by making the best use of naturally rich coastal lands can people hope to eliminate hunger in the world.

People in the United States are unfamiliar with the experience encountered by Europeans in past centuries when the exploitation of natural resources reached a point of diminishing returns. The preservation of forests and wildlife for all people was a major point in the Magna Carta, and the preservation of farm land is an important concern in France. In the United States, green fields are being wiped out by the urban sprawl of housing developments, which will become the slums of the future. Multiple concrete highways choke off the water supply, and the cars that use them pollute the air. Farms are no longer needed in New Jersey and on Long Island because vegetables grow all year round in Florida. But this is accomplished by drying up the Everglades and converting a major incubator of marine life into truck farms. This farm land may in turn be wiped out by shoddy, inefficient housing developments when new areas are opened for exploitation. United States legislators from inland states have little understanding of seacoast problems, and seldom realize the need for marine conservation.

The State of Florida has sold offshore oil leases on its

west coast near Pensacola and Panama City, two areas where the United States Navy has important bases. Although the leased areas cut across sea lanes necessary to the Navy, it has no control over such actions under existing law. It must seek protection from the Army Engineers.

Policies guiding the Army Engineers are not necessarily approved by other groups of people using the water. Cruising yachtsmen object when beautiful sailing grounds are blocked by low automobile bridges sanctioned by the Army Engineers. Bridges had become so widespread by the 1950's that the late Alfred Loomis of *Yachting Magazine* labeled President Eisenhower the "Colossus of Roads." The 1950 Rivers and Harbors Act does not consider berthing areas for pleasure craft eligible for dredging assistance by Army Engineers. Apparently the profit motive is the prime consideration.

In the United States, many interests conflict over the use of the sea, ranging from towns looking for a garbage dump to swimmers, sailors, and fishermen seeking recreation. A convenient way of classifying these varied interests, both in coastal waters and in the ocean, is to divide them into users, observers, and protectors. Typical of the users are the merchant marine, fishermen, offshore oil companies, and seashore vacationers. Among the observers are the marine scientists, the artists and writers, and the millions who admire storm waves on a beach or sunset on the sea's horizon. Protectors include the Coast Guard, who insure safety of life at sea, and the lawmakers and treaty negotiators who protect ocean resources and prevent open conflict.

Some interests span several classifications. The United States Navy, for instance, uses the sea as a means of mobility and protection; it finances, through the Office of Naval

Research, basic scientific observation of the sea; and, with other navies of the world, it protects the traditional freedom of the seas.

Regulation of sea use involves both state and national laws for internal and territorial waters. International law and various treaties control the high seas, along with the seabed. While nations have much to learn about supervising national waters, technical progress continues to create new need for international regulations.

In the United States, agencies concerned with control of inland and coastal waters often represent narrow segments of public interest. They overlap and conflict, and few of them have the broad knowledge and power to avoid disastrous mistakes in the use of valuable resources.

The author E. B. White, reading that the Atomic Energy Commission had authorized the dumping of radioactive waste into the ocean, commented:

I sometimes wonder about these cool assumptions of authority in areas of the sea and sky. The sea doesn't belong to the Atomic Energy Commission, it belongs to me. I am not ready to authorize dumping radioactive waste into it, and I suspect that a lot of other people to whom the sea belongs are not ready to authorize it, either.[1]

The primary need is for broader knowledge of the marine environment to provide the basis for a central agency that can act in the total public interest, not only for today but for future generations. Knowledge must be followed by genuine understanding.

Because the United States lacks a centralized authority responsible for marine problems, the sea is used by different agencies as they see fit. Faced with the problem of disposal of large quantities of World War II nerve gas, the Army investigated various ways of getting rid of it and finally decided to load it aboard obsolete Liberty ships in concrete coffins and to sink them in the deep ocean. At one time this would have been the accepted technique for disposing of wastes, including nuclear material. When news of the nerve-gas disposal became known, the public hue and cry was great enough to halt operations. The Army disposed of the gas in safer ways.

The way the nerve-gas disposal was halted illustrates the complexity of the situation. A legislator called attention to the plan, and the national press gave it wide coverage. Congress reacted, and the Department of Defense stopped the operation. President Nixon referred the matter to the Presidential Scientific Advisory Committee, who in turn called in the National Academy of Sciences. They recommended that the gas be burned and chemically neutralized, with due consideration for atmospheric pollution problems.

If, as recommended by the Presidential Commission on Marine Science, Engineering, and Resources, a new independent federal agency, the National Oceanic and Atmospheric Agency, were formed, this would be a major step toward better national control of the sea, especially if existing agencies with limited or overlapping functions were incorporated and reorganized within NOAA.

At the present time, the United States has two cabinet-level departments for land use and resources: the Departments of Agriculture and the Interior. In contrast, marine agencies are spread throughout the departments, a minority interest within the government. On the state and city level, responsibility for ocean and shoreline activities is divided among various authorities. In many coastal states, agencies charged with marine problems put more effort into mountain parks or trout and deer conservation than into ocean resources.

It is doubtful, however, whether the United States government will take as drastic a step as that recommended by the commission. There are too many internal problems of crime, urban decay, water and air pollution, and failing public transportation to permit the enactment and funding of a major ocean program at this time. It is more probable that the emphasis will be on the immediate problems of the coastal zone and that these problems will be handled by an existing organization, such as the Department of the Interior. However, there is conjecture in government circles that the Interior Department may undergo a major reorganization that would make it a "Department of the Environment," to deal with the problems of the national environment wherever they exist—in the air, on land, or in the water.

Our Nation and the Sea stresses the importance of a "strong, solid base of science and technology." To protect natural resources, planners need better understanding of the interaction of physical and biological forces in the sea. They must know the conditions that exist today, the current data on pollution levels, the size of the marine populations, and the mineral resources. They must have accurate surveys of ocean topography and land-sea boundaries. Until they know what conditions are now, they cannot regulate intelligently or decide whether we are gaining or losing in the struggle for survival. Future generations will depend on the outcome of a battle that man may be losing right now.

At a time when the United States is faced with problems of marine legislation for national waters, it is being forced by growing technology to reconsider its stand on international law for the high seas.

The CEP nations of South America have extended their national rights 200 miles out to sea, far beyond the traditional three miles, or even the twelve-mile fishing limit. Most coastal fishermen would like to extend national jurisdiction as far as possible. Distant-water fishermen, who range thousands of miles after ocean fish, would prefer the old rules of freedom of the seas beyond the three-mile limit. The United States Navy would like to retain a narrow territorial sea because any extension of the marginal strip would close off key straits and limit naval mobility. Offshore oil prospectors who have found indications of petroleum on the continental slope are pressing for national ownership of the seabed beyond the 200-meter depth.

Fishermen and oil prospectors face very different problems and view international law from two opposing points of view. Because most fish range in schools, fishing vessels are designed to follow them over areas of the water. The petroleum companies are tied to oil reservoirs under the seabed and must install fixed equipment in one place. They must have recognized title to seabed property to prevent encroachment from competitors. Oil companies take greater risks, and make greater profits, than the fishing industry.

Oil companies are expert at negotiating for rights where a national government owns title to offshore property. United States, British, Dutch, and French companies have oil properties all over the world. The petroleum industry has led the way in the development of multi-national corporations that spread across national boundaries. Oil companies are suspicious of international control of the ocean floor because there is no successful precedent for such a system. They fear that nations without marine experience, with inflated views of the ocean's wealth, may gain control of any international agency and turn it into an unwieldy bureaucracy. Foolish mismanagement would stifle initiative and delay exploitation of resources in the deep sea.

Fishermen have had experience with broad international control over fisheries, although many treaties are made between only two or three nations actively engaged in fishing. Regulations are set up for large regions of the sea by various commissions and generally prove successful. However, quotas for disputed species, such as the king crab, are usually set up by two nations that understand the immediate problem and have direct control over fishermen. Willing to accept regional control and national treaties about fishing rights, fishermen tend to be suspicious of a large representative body like the United Nations as a lawmaking institution.

About the deep ocean, there seems to be general agreement that some international authority will be necessary to avoid conflict and conserve natural resources. In view of the difficulties most nations have in managing their own internal and territorial waters, development of international regulation will be a long and frustrating task.

Many lawyers believe that the law must be built up gradually from court cases that establish precedents. The International Court of Justice has been set up for this purpose. They feel that such written law as the Geneva Conventions of 1958 is too rigid and artificial for nations that struggle over world power. But lawmakers must not underestimate the rapid growth of technology. Courts move slowly. The old system of precedents is too slow to avoid major conflicts over ocean use or a dangerous waste of resources. Laws agreed upon by national representatives and backed by the weight of world opinion can establish guidelines before events take place. International agreements about Antarctica and peaceful use of outer space are examples of useful legal documents that guide world conduct.

Beyond the conventional concepts of international law, there is an increasing need for a new kind of international regime that will control and regulate the use of the oceans for the benefit of all mankind. The traditional views of national sovereignty do not make sense in the ocean. As Roger Revelle, an ocean scientist who is now head of the Harvard Center for Population Studies, has pointed out, sovereignty in the ocean can lead to great difficulty. Referring to the exploitability clause of the 1958 Geneva Convention, he writes:

> Some international lawyers have interpreted this provision to mean that as deep-water technology advances the coastal states will be able to extend their jurisdiction out to the midpoint of the ocean basins. The long-run consequences of such a division of the ocean into national territories are appalling to contemplate. They would constitute a *reductio ad absurdum* of the concept of nation-states.[2]

Malta's statement to the United Nations, proposing, as a long-term objective, the creation of a special agency "with wide powers to regulate, supervise, and control all activities on or under the oceans and the ocean floor," presents a reasonable approach to international control. Now that the United States and the Soviet Union have reached a tentative accord on the banning of weapons of mass destruction on the sea floor, the way is open for international agreements on the peaceful exploitation of the oceans. It is appropriate that the lead should be taken by the tiny island of Malta, which for centuries has been the hub of the old regime of naval power and control of the Mediterranean.

One of the arguments against international control of the seas is that such control will probably result in an indecisive bureaucracy that will delay action and inhibit capital investment in the ocean. There is no doubt that the problems of organizing an effective international ocean agency are great and that the first steps will be halting; however, the oceans have been there for millennia, and only in the past century have we developed the power both to exploit them and to destroy them. A delay in exploitation in the interest of both today's and future generations seems justified. It may slow the rise in standard of living for the affluent nations that have the developing technology of ocean exploitation; but, on the other hand, it may help to avoid the waste of war.

The Law of the Sea is already a highly developed form of international cooperation. In an age when national seaward boundaries no longer provide geographic isolation, cooperation among nations is vital if we are to eliminate force as an instrument of national policy in the oceans. As world populations expand, the sea's resources will best be exploited and conserved by nations working together. Technical progress is advanced by joint undertakings and free exchange of information. The land has been fought over for generations, and its laws are tangled with tradition. The oceans present a new opportunity to build on good existing law, to develop areas of cooperation, and to experiment with new methods of working together. The great resources of the sea must be used to benefit all men, and its law must serve as an instrument of peace.

DOCUMENTS

DOCUMENT I

Excerpts from the 1958 Geneva Conventions

A. CONVENTION ON THE HIGH SEAS

Adopted by the United Nations Conference on the Law of the Sea, April 29, 1958 (U.N. Doc. A/CONF. 13/L.53)

The States Parties to this Convention,
Desiring to codify the rules of international law relating to the high seas,
Recognizing that the United Nations Conference on the Law of the Sea, held at Geneva from 24 February to 27 April 1958, adopted the following provisions as generally declaratory of established principles of international law,
Have agreed as follows:

Article 1

The term "high seas" means all parts of the sea that are not included in the territorial sea or in the internal waters of a State.

Article 2

The high seas being open to all nations, no State may validly purport to subject any part of them to its sovereignty. Freedom of the high seas is exercised under the conditions laid down by these articles and by the other rules of international law. It comprises, *inter alia,* both for coastal and non-coastal States:

(1) Freedom of navigation;
(2) Freedom of fishing;
(3) Freedom to lay submarine cables and pipelines;
(4) Freedom to fly over the high seas.

These freedoms, and others which are recognized by the general principles of international law, shall be exercised by all States with reasonable regard to the interests of other States in their exercise of the freedom of the high seas.

Article 4

Every State, whether coastal or not, has the right to sail ships under its flag on the high seas.

Article 5

1. Each State shall fix the conditions for the grant of its nationality to ships, for the registration of ships in its territory, and for the right to fly its flag. Ships have the nationality of the State whose flag they are entitled to fly. There must exist a genuine link between the State and the ship; in particular, the State must effectively exercise its jurisdiction and control in administrative, technical and social matters over ships flying its flag.

2. Each State shall issue to ships to which it has granted the right to fly its flag documents to that effect.

Article 23

1. The hot pursuit of a foreign ship may be undertaken when the competent authorities of the coastal State have good reason to believe that the ship has violated the laws and regulations of that State. Such pursuit must be commenced when the foreign ship or one of its boats is within

the internal waters or the territorial sea or the contiguous zone of the pursuing State, and may only be continued outside the territorial sea or the contiguous zone if the pursuit has not been interrupted. . . .

B. CONVENTION ON THE TERRITORIAL SEA AND THE CONTIGUOUS ZONE

Article 1

1. The sovereignty of a State extends, beyond its land territory and its internal waters, to a belt of sea adjacent to its coast, described as the territorial sea.
2. The sovereignty is exercised subject to the provisions of these articles and to other rules of international law.

Article 2

The sovereignty of a coastal State extends to the air space over the territorial sea as well as to its bed and subsoil.

SECTION II. LIMITS OF THE TERRITORIAL SEA

Article 3

Except where otherwise provided in these articles, the normal baseline for measuring the breadth of the territorial sea is the low-water line along the coast as marked on large-scale charts officially recognized by the coastal State.

Article 4

1. In localities where the coastline is deeply indented and cut into or if there is a fringe of islands along the coast in

its immediate vicinity, the method of straight baselines joining appropriate points may be employed in drawing the baseline from which the breadth of the territorial sea is measured.

2. The drawing of such baselines must not depart to any appreciable extent from the general direction of the coast, and the sea areas lying within the lines must be sufficiently closely linked to the land domain to be subject to the régime of internal waters.

Article 14

1. Subject to the provisions of these articles, ships of all States, whether coastal or not, shall enjoy the right of innocent passage through the territorial sea.

2. Passage means navigation through the territorial sea for the purpose either of traversing that sea without entering internal waters, or of proceeding to internal waters, or of making for the high seas from internal waters.

3. Passage includes stopping and anchoring, but only in so far as the same are incidental to ordinary navigation or are rendered necessary by *force majeure* or by distress.

4. Passage is innocent so long as it is not prejudicial to the peace, good order or security of the coastal State. Such passage shall take place in conformity with these articles and with other rules of international law.

5. Passage of foreign fishing vessels shall not be considered innocent if they do not observe such laws and regulations as the coastal State may make and publish in order to prevent these vessels from fishing in the territorial sea.

6. Submarines are required to navigate on the surface and to show their flag.

C. CONVENTION ON THE CONTINENTAL SHELF

Article 1

For the purpose of these articles, the term "continental shelf" is used as referring (*a*) to the seabed and subsoil of the submarine areas adjacent to the coast but outside the area of the territorial sea, to a depth of 200 metres or, beyond that limit, to where the depth of the superjacent waters admits of the exploitation of the natural resources of the said areas; (*b*) to the seabed and subsoil of similar submarine areas adjacent to the coasts of islands.

Article 2

1. The coastal State exercises over the continental shelf sovereign rights for the purpose of exploring it and exploiting its natural resources.
2. The rights referred to in paragraph 1 of this article are exclusive in the sense that if the coastal State does not explore the continental shelf or exploit its natural resources, no one may undertake these activities, or make a claim to the continental shelf, without the express consent of the coastal State.
3. The rights of the coastal State over the continental shelf do not depend on occupation, effective or notional, or on any express proclamation.
4. The natural resources referred to in these articles consist of the mineral and other non-living resources of the seabed and subsoil together with living organisms belonging to sedentary species, that is to say, organisms which, at the harvestable stage, either are immobile on or under the seabed or are unable to move except in constant physical contact with the seabed or the subsoil.

Article 3

The rights of the coastal State over the continental shelf do not affect the legal status of the superjacent waters as high seas, or that of the airspace above those waters.

D. CONVENTION ON FISHING AND CONSERVATION OF THE LIVING RESOURCES OF THE HIGH SEAS

The States Parties to this Convention,

Considering that the development of modern techniques for the exploitation of the living resources of the sea, increasing man's ability to meet the need of the world's expanding population for food, has exposed some of these resources to the danger of being over-exploited,

Considering also that the nature of the problems involved in the conservation of the living resources of the high seas is such that there is a clear necessity that they be solved, whenever possible, on the basis of international co-operation through the concerted action of all the States concerned,

Have agreed as follows:

Article 1

1. All States have the right for their nationals to engage in fishing on the high seas, subject (*a*) to their treaty obligations, (*b*) to the interests and rights of coastal States as provided for in this Convention, and (*c*) to the provisions contained in the following articles concerning conservation of the living resources of the high seas.
2. All States have the duty to adopt, or to co-operate with

other States in adopting, such measures for their respective nationals as may be necessary for the conservation of the living resources of the high seas.

Article 2

As employed in this Convention, the expression "conservation of the living resources of the high seas" means the aggregate of the measures rendering possible the optimum sustainable yield from those resources so as to secure a maximum supply of food and other marine products. Conservation programmes should be formulated with a view to securing in the first place a supply of food for human consumption.

DOCUMENT II

The Truman Proclamation

Proclamation 2667: Policy of the United States with Respect to the Natural Resources of the Subsoil and Sea Bed of the Continental Shelf. September 28, 1945.

By the President of the United States of America a Proclamation:

WHEREAS the Government of the United States of America, aware of the long range world-wide need for new sources of petroleum and other minerals, holds the view that efforts to discover and make available new supplies of these resources should be encouraged; and

WHEREAS its competent experts are of the opinion that such resources underlie many parts of the continental shelf off the coasts of the United States of America, and that with modern technological progress their utilization is already practicable or will become so at an early date; and

WHEREAS recognized jurisdiction over these resources is required in the interest of their conservation and prudent utilization when and as development is undertaken; and

WHEREAS it is the view of the Government of the United States that the exercise of jurisdiction over the natural resources of the subsoil and sea bed of the continental shelf by the contiguous nation is reasonable and just, since the effectiveness of measures to utilize or conserve these resources would be contingent upon cooperation and protection from the shore, since the continental shelf may be regarded as an extension of the land-mass of the coastal nation and thus naturally appurtenant to it, since these resources frequently form a seaward extension of a pool or deposit lying within the territory, and since self-protection compels the coastal nation to keep close watch over activities off its shores which are of the nature necessary for utilization of these resources:

NOW, THEREFORE, I, HARRY S TRUMAN, President of the United States of America, do hereby proclaim the following policy of the United States of America with respect to the natural resources of the subsoil and sea bed of the continental shelf.

Having concern for the urgency of conserving and prudently utilizing its natural resources, the Government of the United States regards the natural resources of the subsoil and sea bed of the continental shelf beneath the high seas but contiguous to the coasts of the United States as appertaining to the United States, subject to its jurisdiction and control. In cases where the continental shelf extends to the shores of another State, or is shared with an adjacent State, the boundary shall be determined by the United States and the State concerned in accordance with equitable principles. The character of high seas of the waters above the continental shelf and the right to their free and unimpeded navigation are in no way affected.

IN WITNESS WHEREOF, I have hereunto set my hand and caused the seal of the United States of America to be affixed.

DONE at the City of Washington this twenty-eighth day of September, in the year of our Lord nineteen hundred and forty-five, and the Independence of the United States of America the one hundred and seventieth.

HARRY S TRUMAN

DOCUMENT III

The Bartlett Bill

Public Law 89-658

AN ACT

To establish a contiguous fishery zone beyond the territorial sea of the United States.

Be it enacted by the Senate and House of Representatives of the United States of America in Congress assembled, That there is established a fisheries zone contiguous to the territorial sea of the United States. The United States will exercise the same exclusive rights in respect to fisheries in the zone as it has in its territorial sea, subject to the continuation of traditional fishing by foreign states within this zone as may be recognized by the United States.

SEC. 2. The fisheries zone has as its inner boundary the outer limits of the territorial sea and as its seaward boundary a line drawn so that each point on the line is nine nautical miles from the nearest point in the inner boundary.

SEC. 3. Whenever the President determines that a portion of the fisheries zone conflicts with the territorial waters or fisheries zone of another country, he may establish a seaward boundary for such portion of the zone in substitution for the seaward boundary described in section 2.

SEC. 4. Nothing in this Act shall be construed as extending the jurisdiction of the States to the natural resources beneath and in the waters within the fisheries zone established by this Act or as diminishing their jurisdiction to such resources beneath and in the waters of the territorial seas of the United States.

Approved October 14, 1966.

Notes

Chapter I
[1] Francis Chichester, *Gypsy Moth Circles the World* (New York: Coward-McCann, 1967), p. 178.
[2] *Ibid.*, p. 179.
[3] Rachel Carson, *The Sea Around Us* (New York: Oxford, 1951), p. 7.
[4] *Ibid.*, p. 12.
[5] See Document I.

Chapter II
[1] *Institutes of Justinian,* trans. John Moyle (Oxford: Clarendon, 1909), p. 36.
[2] See A. P. Higgins & C. Colombos, *International Law of the Sea* (London: Longmans Green, 1954), p. 42.

Chapter III

[1] William Beebe, *Half Mile Down* (New York: Duell, Sloan & Pearce, 1951), p. 132.
[2] See Document II.

Chapter V

[1] See Document III.

Chapter VII

[1] Jacques-Yves Cousteau, "At Home in the Sea," *National Geographic,* April 1964, p. 492.
[2] David Perlman, "An Interview with Commander Scott Carpenter," *Look,* January 21, 1969, p. 78.

Chapter VIII

[1] *New York Times,* October 8, 1969, p. 1.
[2] Arthur J. Goldberg, *Dep't of State Bulletin,* February 12, 1968, p. 193.
[3] Lyndon B. Johnson, "Effective Uses of the Sea," *Weekly Compilation of Presidential Documents,* July 18, 1966, p. 931.
[4] Thomas Pelly, Quoted in "Can the U. S. Parcel Out the Sea-Bed?", *Business Week,* November 11, 1967, p. 67.
[5] Bob Wilson, quoted in *National Oceanographic Association News,* May 1, 1968, p. 2.

Chapter IX

[1] E. B. White, quoted in Garrett Hardin, "Finding Lemonade in Santa Barbara's Oil," *Saturday Review,* May 10, 1969, p. 18.
[2] Roger Revelle, "The Ocean," *Scientific American,* September 1969, p. 56.

Suggested for Further Reading

Anderson, William. *Nautilus-90-North.* Cleveland: World, 1959.
An account of the submarine *Nautilus* and its trip under the ice at the North Pole.

Beebe, William. *Half Mile Down.* New York: Duell, Sloan & Pearce, 1934.
Beebe's first-hand story of building the bathysphere and taking it down on record-breaking dives.

Carson, Rachel. *The Sea Around Us.* New York: Oxford, 1951.
A life history of the sea in all its aspects, told with great beauty and originality.

Chichester, Francis. *Gypsy Moth Circles the World.* New York: Coward-McCann, 1967.
A trip around the world in a sailboat by a crusty Englishman.

Commission on Marine Science, Engineering and Resources. *Our Nation and the Sea.* Washington: Government Printing Office, 1969.
The Commission appointed by the President makes a detailed report on America's use of the sea, laws concerning it, and future possibilities for it.

Cooper, Bryan and Gaskell, T. F. *North Sea Oil: The Great Gamble.* New York: Bobbs Merrill, 1966.
How natural gas was found off the British coast, and an account of the disaster on the oil rig *Sea Gem.*

Cousteau, Jacques-Yves. *Cousteau's World without Sun,* ed. James Dugan. New York: Harper & Row, 1965.
The story of Conshelf II—men living 33 feet beneath the surface of the Red Sea.

Cousteau, Jacques-Yves. "Working for Weeks on the Sea Floor," *National Geographic,* April 1966.
With fine color photographs, Cousteau tells about Conshelf III in the Mediterranean.

Dean, Arthur. "Department Seeks Senate Approval of Conventions on Law of Sea," *Department of State Bulletin,* February 15, 1960.
Reporting to the Senate Foreign Relations Committee, Dean described the 1958 Geneva Conventions simply and completely, in language laymen can understand.

Dugan, James. *Man Under the Sea.* New York: Collier Books, 1965.
A lively discussion of underwater subjects from photography to treasure hunting.

Guberlet, Muriel. *Explorers of the Sea.* New York: Ronald, 1964.
Brief and interesting biographies and descriptions of oceanographers and their expeditions—from Matthew Maury to Willard Bascom.

Hull, Seabrook. *Bountiful Sea.* Englewood Cliffs: Prentice Hall, 1964.
A general look at the sea, full of scientific information, but written for the general reader.

Keach, Donald. "Down to the *Thresher* by Bathyscaphe," *National Geographic,* June 1964.
How the *Trieste* went down to the wrecked submarine and found a twisted hull, part of a shoe, and bits of debris.

Lewis, Flora. *One of Our H-Bombs Is Missing. . .* New York: McGraw-Hill, 1967.
The *Alvin* and the *Aluminaut,* two submersibles, find the missing H-bomb in the waters off Spain.

Pell, Claiborne. *The Challenge of the Seven Seas.* New York: Morrow, 1966.
Senator Pell describes new activities in the sea and future possibilities for the United States in ocean use.

Piccard, Auguste. *Earth, Sky and Sea.* New York: Oxford, 1956.
The great Swiss scientist describes some of his adventures in both the stratosphere and the ocean.

Piccard, Jacques. *Seven Miles Down.* New York: Scribners, 1961.
   The story of the bathyscaphe *Trieste,* and the deepest dive to the Challenger Deep.

Robertson, R. B. *Of Whales and Men.* New York: Knopf, 1961.
   The account of a ship's doctor who traveled to Antarctica with the whaling fleet in 1950 and saw the modern whaling industry at first hand.

Scientific American. "The Ocean," September 1969.
   This entire issue is devoted to the ocean, with articles on geology, marine biology, physical resources, food resources, and man's relation to the sea.

Stenuit, Robert. *Deepest Days.* New York: Coward-McCann, 1966.
   The Belgian diver Stenuit worked with Edwin Link on the early experiments with saturation diving.

Stephens, William. "Sealab II," *Sea Frontiers,* November–December 1965.
   An article about one of the most successful habitats, Sealab II, in a fine little magazine devoted to oceanographic events.

Wagner, Kip. "Drowned Galleons Yield Spanish Gold," *National Geographic,* January 1965, p. 1.
   Wagner's story of diving for the Spanish treasure fleet off Florida, illustrated with fine color photographs.

Index

A page number in italics refers to a diagram on that page.

ABOUKIR, 32
Abu Dhabi, 76
Adriatic Sea, 24
Aegean Sea, 123
Africa, 25, 108, 166; northern coast of, 20; and northern Europe, 23; and Portuguese, 24; as source of slaves, 31; southern part of, 42; offshore oil of, 78; fish farming in, 179
Agriculture, Department of, 183
Alabama: territorial waters of, 70
Alaska, 74, 76, 90, 94, 178
Alaskan North Slope, 76
Alaskan waters: Soviet fishing in, 106
Aleutian Islands, 106
Aleuts, 90
Algeria, 29, 78, 103
Aluminaut, 117, 118, 119, 120, *121*, 123
aluminum, 166
Alvin, 117-120, *121*, 124
Amoy, 29
Amsterdam, 27
anchovetas, 14, 100, 178
Anglo-Norwegian Fisheries Case, 96-99
Antarctic Ocean, 49, 92, 93, 94
Antarctica, 33, 43, 160, 166, 187
aquaculture, 7, 109
Aqua-Lung, 4, 55
Arabian Gulf, 125
arbitration, international, 96-98, 101
archaeology, underwater, 57 58, 123
Arctic Circle, 96
Arctic Ocean, 42, 178
Argentina: and Chile, 33
Arkansas, 109
Army Engineers (U.S.), 66, 155, 181
Ascension, 177
Asherah, *121*, 123
Atlantic Ocean: in period of discovery, 8, 24; privateers and pirates in, 28; eastern, bed of, 82; northeastern, fishing in, 89; seals in, 90; and hurricanes, 176
Atlantic Ridge, 42, 176-177
Atlantis, Isle of Gold, 154-155, 158
Atomic Energy Commission; 182
atomic power plants, 153
Auguste Piccard, 125
Australia, 64, 76, 102-103

Azores, 176, 177

BAHAMAS, 140
Bahrein, 76
Baltic Sea, 23, 25, 33
baralyme, 122
Barbados Oceanographic and Meteorological Experiment, 176
Barents Sea, 41
Bartlett Act (1966), 106, 200-201
Barton, Otis, 48
baseline, *41*, 96-97
bass, striped, 109
bathyscaphe, 49-51, 111-114; *see also Trieste*
bathysphere, 48-49
Beagle, 33
Beaver, *121,* 124
Beebe, William, 47-48, 52
Ben Franklin, *121,* 126-132, 175
bends, 136, 148-149
Benguela Current, 43
Bering Sea, 89, 90, 91
Biafra, 108
Black Sea, 33
Blackbeard, 28-29
blue whale, 93
bluefish, cultured, 109
boat: invention of, ix-x
BOMEX, 176
Bond, Captain George, 146-148, 153
Bonin Trench, 42
Bosporus, 33
Boston, 116
bottom, ocean: contour maps of, 59; living on, 74; *see also* ocean bottom, ocean floor, *and* seabed
Brazil, 64
Brazil Current, 43
Britain, 30, 37, 76, 78, 106; *see also* England
British: and the sea, 17; and whaling, 93
British Admiralty, 34
British colonies, 64
British Isles, 26, 28; *see also* Britain
British Navy, 29, 30, 32
British Petroleum Company, 80
Bucher, Lloyd S., 164
Bureau of Commercial Fisheries, 104, 123, 149, 172
Burke, William, v
Bynkershoek, Cornelis van, 26
Byzantines, 20, 23

CABLES ON SEABED, 12, 29, 123
Caesar, Julius, 22-23

209

INDEX

California, 168; offshore oil wells of, 61–62, 63, 67–69; territorial waters of, 70; offshore phosphorite of, 85; diving near coast of, 122; underwater habitats off coast of, 134, 147; University of, 147
cameras, waterproof, 58
Canada, 90–91, 95, 177, 178
Cannon, Berry, 150–151, 153
canyons in seabed, 42, 111, 177
Cape Cod, 119
Cape Ferrat, 143
Cape Hatteras, 41, 132
Cape Horn, 2–3
Cape of Good Hope, 176
carbon dioxide, 122, 151
Caribbean Sea: diving in, 141
Carolinas: ocean floor off, 131–132
carp, 109
Carpenter, Scott, 134–135, 147, 148–149, 153
Carson, Rachel, 3–4
Carthage, 20, 22
catfish, cultured, 109
CEP nations, 99–101, 185
Challenger Deep, 111, 113, 114, 115
charting the seas, 34
Chichester, Francis, 2–3
Chile, 33, 64, 93, 99, 178
China Sea, 29
Christmas tree (a wellhead), 144
Christy, Francis, v
Civil War (U.S.), 31, 34, 46
clams, 179
coal, 78, 82
Coast Guard (U.S.), 15, 45, 151, 159, 172, 181
Coast Guard and Geodetic Survey, 159
coastal fishermen, 185
coastal waters (U.S.), 181, 182
coastal zone, 171
cobalt, 85
Cobb Seamount, 122, 158–160
cold: divers' feeling of, 138, 139, 140, 141, 144, 149, 151
collision at sea, 35
Colombos, C. J., v
colonies, 30–31, 32
Columbia River, 94
Columbus, Christopher, 24, 34
Commission on Marine Science, Engineering, and Resources, 170–172, 183
computers, 163, 175
Confederate Navy, 46
Congress (U.S.), 101, 162; and offshore oil, 67, 68–69; and fishery zone, 106; and study of sea, 170; and disposal of nerve gas, 183
conservation: of sea's resources, 39, 180, 188; of fish, 88, 89, 98, 105, 106, 110; of fur seals 89–91; of whales, 89, 92–94
Conshelf, 120, 142–144, *145,* 153
Constantinople, 25
Construction Differential Subsidy Program, 105
contiguous zone, *41*
continental shelf: (U.S.) once exposed, 6; control of, 14, 64, 70, 155, 157; and Geneva Conference (1958), 40–41, 157, 165; described, 40–41; diagram of, *41;* width of, 41, 170–171, 172; oil found in, 61, 135 (*see also* offshore oil); Truman's proclamation on, 62–64; under North Sea, 79; as source of oyster shells, 83; minerals on, 87; and CEP countries, 99; surveyed with submersible, 125; broadcasting from, 156–157; importance of outer limit of, 168; nuclear power plant on, 171
continental slope, 40, *41,* 42, 106, 185
Convention for the Preservation and Protection of Fur Seals, 90–91
Convention on Fishing and Conservation of the Living Resources of the High Seas, 98, 196–197
Convention on the Continental Shelf, 64, 157, 168, 195–196
Convention on the High Seas, 36, 191–193
Convention on the Territorial Sea and the Contiguous Zone, 193–194
Cook, Captain, 32
Copenhagen, 32
copper, 85, 86
coral reefs, 154–156
Corlieu, Louis de, 53
Cousteau, Jacques-Yves, 53–55, 58, 120, 122, 133–135, 141–143, *145,* 146, 153
crabs, 15, 104, 148, 186
Crete, 21
Cuba, x
Cubmarines, 117, 123
cultured fish, 109
currents of ocean, 42–43, 178–179
CURV, 118

DAKAR, 51
Dardanelles, 33
Darwin, Charles, 33
DDT, 174
deck chamber: for decompression, 139–140, 141, 150; for compression, 149
decompression, 136–138, 139–140, 141

INDEX

decompression chambers, 137, 138, 140, 141, 146
decompression tables, 136
Deep Cabin, 142, 143
deep ocean: floor of, 40, 165; shape of vessels for, 113; control of, 171, 172, 186; dumping in, 183
Deep Quest, 121, 123-124
Deep Sumbmergence Rescue Vehicle, 124
Deep Submergence Search Vehicle, 124
Deepstar-2,000, 122, 159
Deepstar-4,000, 5, *121,* 122
Deepstar-20,000, 121
Defense, Department of, 162, 183
Denmark, 33, 78-79
diamonds, 82-83
diesel submarines, 160
Disarmament Conference (1969), 163
divers, 44-45, 53, 56, 115, 117, 136
diving: with scuba gear, 52-53, 55
diving apparatus, 44
diving clubs, 56-57, 58
Diving Saucer, 120, *121,* 122, 143, 144
diving suits, 133, 149
diving techniques, 44
Dogger Bank, 41
Doldrums, 176
dolphins, 179
Donald Duck effect, 138, 141
Drake, Sir Francis, 27, 28
DSRV, 124
DSSV, 124
Dubia, 76
Duke University, 152
Dumas, Frédéric, 55
Dutch, 27, 93; *see also* Holland

EARTH: ROTATION OF, 42, 43
echo sounder, 98
ecological balance, 180
Ecuador, 64, 99, 100-101
Edward III of England, 24
Egypt, 19, 37-38, 103
Eisenhower, President, 69, 181
Elath, 37
Electric Boat Company, 46
Elizabeth I of England, 25
Elk River, 150
Emery, K. O., 86
England: and ancient Romans, 22; as "King of the Seas," 24; fishing rights near, 24; and Spanish shipping, 25; territorial limit of, 26, 27, 28; Dutch maritime wars with, 28; and Barbary pirates, 29; and United States, 29-30; and France, 30; and traffic in slaves, 30-31; smuggling into, 31; and straits, 33; and charting of seas, 34; divides North Sea floor with Norway, 79; and whaling, 93; arbitrates fisheries dispute, 96-97; *see also* Britain *and* British
English Channel, 23, 25
ensign, 29
Environment, Department of the, 184
Environmental Science Services Administration, 172, 176
epoxy resin, 113
Equatorial Current, 43
Eskimos, 90
ESSA, 176
euphoria of the deep, 138
Europe: northern, 23, 77-78; and Gulf Stream, 43, 176; conservation in, 180
Everglades, 180
Exploring Expedition of 1838-42, 33-34

FACTORY SHIPS, 11
FAO, 99
Far East: Dutch trade with, 27
farming the sea, 109, 135
fathometer, 129, 177
fertilizers, 83-85, 178
Fiji Islands, 34
fin whale, 93
Fire Standards for Passenger Vessels, 35
fish: as food, 19, 21, 106-107; as wild and as property, 23; of the sea, Grotius on, 26; language of, 59; sedentary, 64; raising, feeding, herding, 109-110, 143, 179; caged, 148; of deep ocean, 178; of open ocean, 179; schooling, 179
Fish and Wildlife Service, 90, 158
fish meal, 6, 100, 105, 107
fish oil, 105
fish protein concentrate, 107-108
fisheries, ocean: of North Sea, 27; Truman's proclamation on, 62; regulation of, 88, 89, 95-96, 98, 186; conservation of 88, 89, 98; and the law, 96, 98, 99; estimating resources of, 98-99; distant-water fleets for, 101-103; total world catch of, 104; in United States economy, 106; as source of food for expanding populations, 107; treaties regulating, 186
fishery zones, 106, 110
fishing: to the Greeks, 21; sonar used in, 59; importance of, 60
fishing boats, 37
fishing industry, 185

fishing rights: in North Pacific, 15-16; near England, 24; based on historic usage, 38-39; based on need, 39; extended, 63, 64; acquired by treaty, 88, 186; of Norway, 98; and territorial waters, 99; of Peru, 100; and Peru and Ecuador, 100-101; to cultured fish, 110
fishing waters: pollution of, 175
fishways for salmon, 95
flag, ship's, 29; of convenience, 36
fleets, fishing: large, 101-103, 104, 105; distant-water, 110
Flip, 176
Floating Instrument Platform, 176
Florida, 13-14, 70, 126, 130; coast of, 154-155, 158; as source of vegetables, 180; offshore oil of, 180-181
fluke, 106
FNRS, 51, 111, 112
Fonds National Belge de la Recherche Scientifique, 51
food from sea, 6, 103, 133, 179
Food and Agricultural Organization (U.N.), 99
Food and Drug Administration, 107
FPC, 107-108
France, 29, 30, 32, 78, 180
Fraser River, 94
freedom of the seas, 22, 25, 26, 28, 30, 119, 185; an exception to, 32; in 19th century, 32; and 1958 Geneva Conference, 36; preservation of, 39; and U.S. Navy, 182
freeze-dried food, 128, 147
French Navy, 51, 111
French Riviera, 139
frogmen, 56, 117
Fuller, William, v
Fulton, Robert, 46
fur seals, 89-91

GAGNAN, EMILE, 54
Galápagos Islands, 33
gas, natural, 77, 78, 79, 80, 83
gasoline, 50
General Assembly (U.N.), xi, 165
General Dynamics, 123
General Electric, 151
Geneva, 125, 163
Geneva Conference of 1958, 36-37, 64, 79, 110, 163, 165; defines continental shelf, 40-41; on Law of the Sea, 98
Geneva Conference of 1960, 37
Geneva Conventions of 1958, 186, 187, 191-197

Genoa, 24
Germany, 32, 78, 112-113
Geyer, Leo, v
Gibraltar, 33
glaciers, 41, 42
Glomar Challenger, 82
goggle diving, 53
gold, 86, 166
Goldberg, Arthur J., 165
Grand Banks, 104
Grand Capri Republic, 154-155, 158
Great Lakes, 171
Great Stirrup Cay, 140
Greece, 57
Greeks and the sea, 7-8, 21
Greenland, 8, 25
Groningen, 78
Grotius, Hugo, 25-26, 88
grouper, 141
Grumman Aerospace Corporation, 126
Guam, 111, 161
Gulf of Aqaba, 37, 38
Gulf of Mexico, 74, 81, 82
Gulf Stream, 43, 130, 131, 132, 175-176
Gulf Stream Drift Mission, 129-132, 175
gypsum, 84
Gypsy Moth, 2

HABITATS, UNDERWATER, 44, 55, 133, 134, 135, 139, 140-144, 146-147, 149-150; on ocean floor, 11, 18; for lobsters, 110; served by submersibles, 115; and seamounts, 158-159, 160; for exploration, 177
Hague, the, 37, 89
Haldane, John Scott, 136
halibut, 95, 104
Halley, Edmund, 44
Harvard Center for Population Studies, 187
Hawaii, 123, 158
Hawkins, Sir John, 27
helium: used in gaseous mixture breathed by divers, 54, 146, 147, 150, 151; increases feeling of cold, 138, 139, 140, 141, 144; affects voice, 138, 141, 142, 144
herding fish, 109, 133, 135
herring, 27, 28, 95
Heyerdahl, Thor, 2
high seas, 158, 164, 165, 184-185; Geneva Convention on, 191-193
high-frequency scanning, 161
Holland: and Portugal, 25; maritime wars of, with England, 28; and Barbary pirates, 29; gas discovered in, 78, 79; broadcasting in, 156, 157; *see also* Dutch

INDEX

Holland, John, 45, 46
Honduras, 64
Hong Kong, 29
Howe Sound, 125
Hudson Canyon, 42
Humboldt Current, 43, 175, 178
hurricanes, 176
hydrodynamics, 179
hydrogen bombs, 116–118
hydrophone, 59

ICE AGE, 41
Iceland, 42, 83, 99, 176, 177
Ickes, Harold L., 65, 67–68, 69
India 19
Indian Ocean, 25, 37, 176
Indians, American, 31
Indonesia, 99
innocent passage, 36–37
Institutes of Justinian, 21–22
intelligence ships, 164–165
Interior, Department of the, 66–67, 68, 151, 155, 172, 183, 184
intermediate zone, 171
internal waters, *41*
International Bureau for Whaling Statistics, 92–93
International Conventions for the Safety of Life at Sea, 35
International Court of Justice, 79, 186
International Decade of Ocean Exploration, 169
international law, 20, 29; and slave trade, 31; for control of sea traffic, 35; of the sea, 36, 182, 184, 185 (*see also* Law of the Sea); and offshore oil, 60; and floor of continental shelf, 64; on fisheries, 88–89; and 1958 Geneva Conference, 98; and conservation of fish, 110
International Load Line Convention, 36
International Pacific Salmon Fisheries Commission, 95
International Rules for Prevention of Collision at Sea, 35
International Whaling Commission, 93
Iran, 76
Iraq, 76
iron, 82
islands, 154–155, 168–169, 177
Israel, 37–38
Italians, 55–56

JAPAN: AND KUROSHIO CURRENT, 43; and Australia, 64; seabed mines of, 82; and fur seals, 90–91; whale meat in diet of, 92; and whaling, 93–94; and Pacific salmon, 95; competition of, in fishing, 99; fishing fleets of, 101, 103; and fish culture, 109; submersibles of, 125; fish farming in, 179
Johnson, Douglas, M., v
Johnson, President Lyndon B., 167
Jones, John Paul, 28

KING CRAB, 15, 186
King of the Seas, 24
Kuroshio, 125
Kuroshio Current, 43, 175
Kuwait, 76

LAKE, SIMON, 46–47
Lake Geneva, 125
Lake Maracaibo, 61
Langevin, Paul, 59
Law: of fisheries, 110; based on court cases, 186; *see also* international law *and* Law of the Sea
Law of the Sea, 13, 39, 160, 188; international conferences on, 16; uncodified, 16; 1958 conference on, 17; present attitude toward, 17–18; history of, 18; always international, 20; Geneva conference on, 36, 165; and continental shelf, 40, 79; codified (1958), 98; and fishing, 110; and conservation, 110; and ownership of ocean, 110
League of Nations, 16
Lebanon, 19
Le Prieur, Yves, 53, 54
Liberia, 36
Libya, 78
life-support systems, 122, 126, 127–128, 146
lime from oyster shells, 83
Lindbergh, Charles A., 140
Lindbergh, Jon, 140–141
Link, Edwin, 123, 138–140, 141, *145,* 146, 153
loading regulations, 36
lobsters, 109, 110
Long Island, 180
Loomis, Alfred, 181
Louisiana, 61, 66, 70, 72, 83, 109
Lulu, 120

MAGNA CARTA, 180
magnetic detectors, 116
Malta, 86, 165, 166, 167, 168, 169, 187–188

INDEX

mammals of the sea, 4, 179
man: returns to sea, 4; as polluter of sea, 5; see also pollution
manganese, 85, 86, 87, 119, 166
maps, marine, 34
Mariana Trench, 42, 111, 113
mariculture, 7
marine biology, 175, 179
marine botany, 175
marine geology, 175
marine life, 179
maritime law, 35; see also Law of the Sea
Marseille, 142
Maury, Matthew Fontaine, 34-35
McDougal, Myres, v
Mediterranean Sea: in Ancient World, 8, 20; Rome and Carthage in, 11, 20, 22; pirates in, 22-23, 29; and Romans, 22-23; in age of discovery, 24-25; and Black Sea, 33; archaeological diving in, 57; regulation of fishing in, 89; *Trieste* submerges in, 112; diving in, 57, 122, 139; underwater living in, 143; control of, 188
merchant marine, 181
Miami, 141
mid-Atlantic ridge, 42, 176-177
Middle Eastern oil, 76-77
migration of fish, 94-95
military installations on seabed, 165
milk protein, 108
milkfish, 109
Mine Defense Laboratory, 149
minerals from seabed, 85-87, 156, 166
mini-subs, 115; see also submersibles
Mississippi, 109
mistral, 139
moon, 42
Moslems, 23
mother-of-pearl, 83
mountains, underwater, 42, 158, 176-177
movies, underwater, 58
mullet, 109
museums, aquatic, 135

NAPOLEON, 32
NASA, 131, 151
National Academy of Sciences, 183
National Oceanic and Atmospheric Agency, 172, 183
National Oceanographic Data Center, 172
national sovereignty, 187
natural gas, 77, 78, 79, 80, 83
natural law, 22
Nautilus, 160, 177

Naval Observatory and Hydrographic Office, 34
Naval Oceanographic Office, 82, 131
naval warfare, 9-10, 24, 28
navigation: freedom of, 33; obstructions to, 175
Navigation Acts (British), 30
navy: first, 21; Roman, 22, 23; see also British Navy, French Navy, United States Navy
Navy Board of Inquiry, 165
Navy Depot of Charts and Instruments, 34
Near East, 23, 76-77
Nekton, 124
Nelson, Admiral Horatio, 32
Netherlands, 156-157; see also Dutch and Holland
New Amsterdam, 28
New England, 30, 158
New Jersey, 180
New York, 28
Newfoundland, 104
nickel, 85, 86
Nigeria, 74, 108
Nile, 19
nitrates, 178
nitrogen, 136, 137-138
nitrogen narcosis, 138, 146
Nixon, President, 183
NOAA, 172, 183
Norsemen, 8
North Africa, 29
North American, 30
North American Rockwell, 124
North Atlantic, 25, 43
North Korea, 164, 165
North Pacific: fishing rights in, 15-16
North Pacific Fisheries Convention, 95
North Pole, 49, 178
North Sea, 41, 156, 158, 172; herring fishery in, 27; rough weather in, 74; natural gas discovered in, 77, 78, 80; hazardous, 81; conference on fishing in, 89
North Seas Fishing Convention, 89
northern hemisphere: winds in, 43
Norway, 79, 92, 93, 96-98
Nova Scotia, 130
NR I, 124
nuclear mines, 163
nuclear missiles, 160-162
nuclear submarines, 116, 124, 160-162, 177
nuclear weapons, 163, 164

OCEAN: MAPPING AND CHARTING OF, 34; resources of, 39; deep, 82, 158; economic

INDEX

exploitation of, 186, 187, 188; control of, 187; *see also* sea
ocean bottom, 40, 42, 43, 47; *see also* ocean floor *and* seabed
ocean floor: listening posts on, 162–163; wealth of, 166, 167; national ownership of, 167–168; control of, 172, 185, 187; geology of, 175; under Polar ice cap, 177–178; food from, 178
Oceanographer, 159
oceanography, 34, 59, 175
Office of Naval Research, 181–182
offshore oil, 66–77, 82, 83, 174–175; discovery of, 60; drilling for, 72, 75; off Florida's west coast, 180–181; and territorial limit, 185
oil: in seabed, 12, 39, 59, 66, 156, 166; importance of, 60; underwater storage of, 125; per capita use of, 174
oil industry, 153, 168, 185
oil rigs, offshore, 71, 72, 73, 81
oil spills, 174–175
oil-well installations: underwater work on, 123, 124, 135, 143, 144
optimum sustainable yield, 98
Oslo, 93
Our Nation and the Sea, 170, 184
Outer Continental Shelf Lands Act (1953), 70, 155–156
ownership: of ocean, 21, 24–26, 38–39, 52; of seabed, 60, 62, 63, 64, 87, 185; of continental shelf, 64, 65, 70; of fish, 104, 110
oxygen, 122, *127,* 137, 138, 139
oysters, 6, 64, 83, 109, 179

PACIFIC OCEAN: IN WORLD WAR II, 8, 10; explored by Cook, 32; increased knowledge of 34; arcs of islands in, 42; currents in, 43; near Alaska, hazardous for oil rigs, 81; regulation of fishing in, 89, 95; deepest point in, 111; Humboldt Current in, 178
Pacific Salmon Convention, 95
Palomares (Spain), 116, 124
Panama, 36
Panama City (Florida), 181
papacy, 25
Paris; Treaty of (1763), 30
Parliament, British, 30
pearl beds, 43
pearl fisheries, 64
Pell, Senator Claiborne, 167, 172
Pelly, Thomas, 168
Pennsylvania, University of, 123, 151

Pensacola, 181
Perry company, 123
Persian Gulf, 64, 76, 172
Peru, 64, 99, 100–101, 107, 178
Peter the Great Bay, 99
petroleum, 77, *84; see also* natural gas *and* oil
Philadelphia General Hospital, 149
Philippine Trench, 42
Philippines, 43
Phoenicians, 7, 19–21
phosphates, 119, 178
phosphorite, 83–85, 87
phosphorus, 85
photography, underwater, 55, 56, 58
Physical Geography of the Sea, 34
Piccard, Auguste, 49, 51, 52, 55, 111
Piccard, Jacques, xii, 51, 111, 112, 113–114, 125, 126, 129–130, 175–176
piracy: and Geneva Conference, 36
pirate radio-television stations, 156–158
pirates, 21, 22–23, 28–29
Pisa, 24
Pisces, 121, 125–126
plankton, 7, 148
plants: as source of all food, 178
Plimsoll mark, 36
Polar ice cap, 177
Polaris missile, 161
poles of earth, 42
pollution: of sea, 32, 167, 174–175; of rivers, 95; of air, 180, 183
Pompey, General, 23
ponds, artificial, 109
population, expanding, 107, 110
porgie, 106
porpoise, trained, 147
Portugal, 24, 25, 31
Poseidon missile, 161
Prado, Arvid, 86, 165–166
Presidential Commission on Marine Science, Engineering, and Resources, 183
Presidential Scientific Advisory Committee, 183
pressure, deep-sea, 133, 134
Pribilof Islands, 90
Privateer, 132
privateers, 28
profit motive, 181
Prohibition, 31–32
Project Sea Use, 158–160
Projects Tektite, *145,* 151–152, *152*
Protector, H.M.S., 3
protein, 6, 106–107, 108, 125
Pueblo, 164–165

RADAR, 161
radioactive waste, 36, 165, 182
rainbow trout, 109
Ray, Louis, 154
Red Sea, 37, 86, 134, 142, 143
reefs, artificial, 109–110, 179–180
regional commissions, 89
regional conferences, 89, 110
registration of ships, 29, 36
rescues from downed submarines, 135
Revelle, Roger, 187
Revolutionary War, 29–30, 46
Rhode Island, 167, 172; University of, 109
Rhodes, 20
Rhodian law, 20
Rickover, Admiral H. G., 160
right whale, 92
rights of the high seas, 64
rivers: in the ocean, 41, 175–176
Rivers and Harbors Act (1950), 181
Roaring Forties, 2
Roman law, 94, 110
Rome, ancient, 20, 21–22
Roosevelt, Franklin D., 65, 67
Roosevelt, James, 65
rorqual whale, 92
Royal Dutch Navy and Police, 156
royalty on oil, 66
Russia, x, 33, 90, 93–94, 114; *see also* Soviet Union

SAFETY AT SEA, 35
Salfaniya, 76
Salia, Dr. Saul, 109
salinity of sea, 179
salmon: Pacific, 94–95, 104; cultured, 109
salt domes, 82, 83, *84,* 166
salt fish, 21
salvage, 14, 44–45, 47, 125–126, 135, 148
San Juan, 101
Santa Barbara Channel, 12, 72
saturation diving, 137, 149, 153
Saudi Arabia, 76
Scandinavia, 25, 27, 37
scanning sonar, 129
Schaeffers, Edward, v
schooling fish, 179, 185
Scientific Advisory Committee, 183
scientific knowledge, 98, 110
scorpion fish, 148
Scotland, 27, 161
Scott, Anthony, v
Scripps Canyon, 148

Scripps Institution of Oceanography, 147
scuba gear, 52–54, 56
SDC, 138–139, 140, 141, 142, *145*
sea: to whom it belongs, x–xii, 1, 13, 18; riches of, xi, xii, 60, 188; human feelings about, 1–2, 3, 5, 8; and new technology, 2; man's return to, 4; polluted by man, 5; as source of food for man, 6–7 (*see also* fish *and* fisheries); used in war, 9–10, 24; becoming crowded, 13; drawing boundaries on, 13; ownership of, 13–14, 18 (*see also* ownership); control of, 14–16; as pathway to knowledge, 33–34; intelligent use of, 175; classifying those interested in, 181–182; *see also* ocean
sea farming, 7, 109, 135
Sea Gem, 80–81
Sea Grant Program, 172
sea law, 14–15, 22; *see also* Law of the Sea
Sea of Japan, 158
sea plants: harvesting, 133, 135
sea power, 21, 32
sea transportation, 8, 9, 11
seabed: and law, 13; control of, 14, 160, 167; strewn with mineral chunks, 85–87; canyons, in, 111; ecology of, 152; military installations on, 163, 166, 167–168, 187; of deep ocean, 165–166; covered with nutrients, 178; *see also* ocean bottom *and* ocean floor
seabirds, 178
Seacliff, 121
seacoast, 180, 181
Seafreeze Atlantic, 105
Sealab, *145,* 146, 147–151, 153
seamounts, 42, 122, 158–160
seashells, 83
Security Council (U.N.), xi
sediment on seabed, 5, 41
seismic studies, 59, 125
Selden, John, 26
Self-Contained Underwater Breathing Apparatus, 52
Senegal, 103, 166
Sharm El Sheik, 37, 38
Shelf Diver, 121, 123
shellfish, 179
shipbuilding, 19
shipping: protection of, 38
shrimp, 102, 104, 109
Siberia, 177, 178
Sidon, 21
Siebe, Augustus, 44
silver, 86

INDEX

Sinai Peninsula, 37, 38
Six Days' War, 38
skin divers, 56
slates, underwater, 58
slavery, 30-31, 36
smuggling, 31-32, 36
Society Islands, 33-34
SOLAS, 35
solid-fuel missile, 161
sonar, 58-59, 92, 98, 116, 129, 161, 162-163
SOSUS, 162-163
Soucoupe Plongeante, 120
sound fathometer, 177
South America, 33, 37, 41, 43, 178
South Atlantic, 43, 176
South Pacific, 2, 33
Southeast Asia, 82
southern hemisphere, 43
sovereignty over sea, 21
Soviet Union: and territorial waters, 37, 99; oil supply of, 80; islands of, in Bering Sea, 91; Pacific coast of, 99; fishing boats and fleets of, 101-103, 106; submarine fleet of, 162; and Disarmament Conference of 1969, 163; and U. N. Assembly, 164; and United States, 187; *see also* Russia
soybeans, 107
Spain, 14, 19, 25, 116, 118, 119, 161, 166
sperm whale, 91
SPID, 140-141, *145*
sponge fisheries, 43
Stamp Act, 30
Star, 121, 123
starfish, 148
Starfish House, 142-143
states (of U. S.), 14, 61, 62, 63, 66
states' rights, 65-66, 69, 70
steamships, 32
Stenuit, Robert, 139-141
St. Helena, 177
St. John, 151
Strait of Gibraltar, 33
Strait of Magellan, 33
Strait of Tiran, 37, 38
straits, 32, 33, 37
stratosphere, 49
stratospheric balloon, 49, 51
Stratton report, 170
striped bass, 109
strobe light, 58
submarines: in World War I, 10; nuclear, 10, 116, 124, 160-162, 177; small, 11; in innocent passage, 37; early, 45-47; present, 47; escaping from downed, 52; detection of, 58-59; military, 114, 115, 127; sounds made by, 163; not banned in 1969, 163
submerged lands, 61-63, 66-68, 69, 70, 74-76
Submerged Lands Act (1953), 70
submerged portable inflatable dwelling, 140
submersible decompression chamber, 138-139, 140, 141, 153
submersibles: Cousteau's, 55, 120, *121,* 122, 143, 144; future, 74, 135; used in underwater searches, 115-118; maximum depths of, *121;* new, 122, 123-126; for exploration, 122, 177; and seamounts, 160
Suez Canal, 76, 77, 78
sulphur, 83, *84*
sun, 42
super-insulated tanks, 128
Survey Service and Hydrographic Department of British Admiralty, 34
surveys, underwater, 175, 177, 178
Sweden, 108
swim fins, 53
Switzerland, 125, 176
swordfish, 119, 131
Syria, 19

TEACH, EDWARD, 28-29
technology: new, and the sea, 2, 10, 11, 18; in underwater exploration, 45-46; revolution in (1960's), 115; growth of, 186
Tektite, *145,* 151-152
telephone, underwater, 129
television, underwater, 116, 129, 147, 148, 150, 153
temperature of sea water, 179
territorial limit, 32, 66, 155, 156, 158, 164, 165
territorial sea *41, 96, 97; see also* territorial waters
territorial waters: early ideas of, 26-27; of United States, violated, 30; extension of, for special purposes, 32; narrow, favored by United States and Britain, 32, 106; and Geneva Conferences (1958, 1960), 36-37; United States' view of, 37; Soviet view of, 37; various widths of, 37, 64, 185; as extension of land, 38; as buffer zone, 38; of the states (U.S.), 70; fur seals breed in, 92; Norway's, 97-98; and fishing rights, 99; Indonesia's, 99; of CEP countries, 99-100; distinguished from fishery zone, 106; of Florida, 155; laws concerning, 182
Texas, 69, 70, 77
three-mile limit, 14, 155, 156, 158
Thresher, 116, 124

throat microphone, 58
tidal pools, 179-180
tidal wetlands, 180
tidelands oil controversy, 66
tides, 42
tin, 82
Titanic, 35
Trafalgar, 32
traffic at sea: control of, 35
transfer capsule, 149, 150
Transportation, Department of, 172
trawl nets, 27
treaties: on the sea, x, 182; on fishing rights, 105; on fisheries, 110
trenches in ocean floor, 42, 111, 113, 177
Trieste, 51, 111-114, 116, *121*
Tristan da Cunha, 176-177
Triumph Reef, 154
tropics, 42
trout, 109
Truman, President, 62-63, 64, 68-69
Truman Proclamation, 62-64, 198-199
tuna, 100-101, 104
Tunisia, 43
Turkey, 25, 33
TV Nordzee, 156-157
twelve-mile fishery zone, 106
twelve-mile limit, *41,* 164
Twenty Thousand Leagues Under the Sea, 58, 133
Tyre, 21

UNDERSEA WARFARE CENTER, 149
underwater cameras, 116
underwater silos, 135
underwater storage tanks, 135
underwater technology, 45-47
underwater vehicles, 44, 48, 49, 51, 52, 115, 159, 175; Navy program for, 124
underwater warfare, 55
underwater world, 44, 55
United Nations: and the sea, x, 16-17; Security Council of, xi; Assembly of, xi, 164; and 1958 Geneva Conference, 36, 170; as peacemaker, 38; World Court of Justice of, 96-97; and control of seabed, 160, 169-170, 172; and Geneva Disarmament Conference (1969), 163; report to, on *Pueblo* case, 164-165; and deep seabed, 165-166; and Malta Proposal, 169; permanent seabed committee of, 169; as lawmaker, 186
United States: and territorial waters, 27, 37, 106; and Barbary pirates, 29; and England, 29-30; and slave traffic, 30-31; smuggling into, 31-32; sends out Exploring Expedition, 33-34; and Gulf Stream, 43; oil supply of, 80; and protection of fur seals, 90; islands of, in Bering Sea, 91; and conservation of Pacific fish, 94, 95-96; and CEP countries, 100-101; rank of, in ocean fishing, 103-105; and foreign fishing fleets off coast, 105-106; sends FPC to Biafra, 108; prevents building of new islands, 155; and Cobb Seamount, 160; sea defenses of, 160-163, 169; tracks Soviet submarines, 162; and military use of seabed, 163; and Disarmament Conference of 1969, 163; and U. N. Assembly, 164; intelligence ships of, 164-165; and Malta Proposal, 167-168; and fish farming, 179-180; and ecological balance, 180; and use of the sea, 181; marine problems of, 182; and Soviet Union, 187
United States Army, 182-183
United States Navy: and pirates, 29; in War of 1812, 30; buys *Trieste I,* 51, 112, 113; and offshore oil, 61-62, 68; finds the *Thresher,* 116; retrieves submerged hydrogen bomb, 116-118, 120, 122; program of, for underwater vehicles, 124; divers of, 135, 146, 147, 148; publishes decompression tables, 136; works on underwater habitats, 140, *145,* 146-151, 153; develops missiles, 161; and salt domes, 166; plans of, for deep seabed, 167; and relation to sea, 182; and territorial limit, 185
United States versus California, 69
USSR, *see* Russia *and* Soviet Union
upwelling currents, 178-179

VANCOUVER, 125-126
Venezuela, 61, 74
Venice, 24
Verne, Jules, 44, 58, 133
vest, insulating, 144
Virgin Islands, 151
Vladivostok, 99
volcanic eruptions, 177
volcanic islands, new, 177
volcanic peaks in deep ocean, 158

WAGNER, KIP, 14
Walsh, Donald, 111, 113-114
war at sea, 9-10, 24, 28
War for Independence, 30

INDEX

War of 1812, 30
warning system (against submarines), 162–163
Washington, State of, 158, 168
watches, waterproof, 58
water ballast, 46
water of sea: movement of, 42–43
water-pressure gauges, 129
Weather Bureau, 172
Wells, James, v
West Indies, 30, 31, 43, 158
Westinghouse Company, 122
wetlands, 180
whales, 24, 89, 91–94, 179
White, E. B., 182
wildlife: preservation of, 180
Wilkes, Lieutenant Charles, 33
Williamson, John, 58

Wilson, Bob, 168
winds, 42–43
Wonsan, 164
Woods Hole, 120
Woods Hole Oceanographic Institution, 86, 119
World Court, 16
World War I, 47, 59
World War II, 47, 53, 55, 56, 58, 60, 68, 103, 133, 182

YACHTING, 181
Yachting Magazine, 181
Yomiuri, 125

ZINC, 86
zirconium, 85

Picture Credits

From The Bettman Archive: pictures 1 through 6, 10, 21, 26 through 31.

From United Press International (UPI): pictures 7, 13, 19, 20, 22 through 25, 33 through 38, 41 through 46, 49, 50, 52, 53, 54, 60, 63, 64, and 66.

From Wide World Photos: pictures 8, 15 through 18, 32, 39, 40, 55 through 59, 61, 62, and 65.

From U.S. Naval Institute: pictures 9 and 11.

From United Nations: picture 12.

From Texas Gas Transmission Corp.: picture 14.

From Grumman Aerospace Corp.: pictures 47 and 48.

U.S. Coast Guard photo from UPI: picture 51.

15289

551.46
SCO   15289

AUTHOR
Scott, Frances
TITLE
Exploring ocean frontiers

DATE DUE	BORROWER'S NAME
	C. Kret
MAR 19 '9_	R. Lucas
	C. Krust
DEC 21 '95	C. Kreist

551.46
SCO   15289

Scott, Frances
    Exploring ocean frontiers

St. Mary's H. S. Library
South Amboy, N. J.